THE KLUWER INTERNATIONAL SERIES IN ENGINEERING AND COMPUTER SCIENCE

MODELING OF ELECTRICAL OVERSTRESS IN INTEGRATED CIRCUITS

Carlos H. Díaz
Integrated Circuits Business Division, R&D Center
Hewlett-Packard Company
Palo Alto, California, USA

■

Sung-Mo (Steve) Kang
Coordinated Science Laboratory
University of Illinois
Urbana, Illinois, USA

■

Charvaka Duvvury
Semiconductor Process and Design Center
Texas Instruments Incorporated
Dallas, Texas, USA

KLUWER ACADEMIC PUBLISHERS
Boston/London/Dordrecht

Distributors for North America:
Kluwer Academic Publishers
101 Philip Drive
Assinippi Park
Norwell, Massachusetts 02061 USA

Distributors for all other countries:
Kluwer Academic Publishers Group
Distribution Centre
Post Office Box 322
3300 AH Dordrecht, THE NETHERLANDS

Library of Congress Cataloging-in-Publication Data

A C.I.P. Catalogue record for this book is available from the Library of Congress.

Printed on acid-free paper.

Printed in the United States of America

To Rossy, Carlitos, Julia, Juan,
Mia, Jennifer, Jeffry,
Vasu, Rolan.

CONTENTS

LIST OF FIGURES

Chapter 3

Chapter 4

Chapter 5

Chapter 6

Chapter 7

Chapter 8

LIST OF TABLES

PREFACE

Electrical Overstress (EOS) and Electrostatic Discharge (ESD) pose one of the most dominant threats to integrated circuits (ICs). These reliability concerns are becoming more serious with the downward scaling of device feature sizes. This book presents a comprehensive analysis of EOS/ESD-related failures in I/O protection devices in integrated circuits.

The design of I/O protection circuits has been done in a *hit-or-miss* way due to the lack of systematic analysis tools and concrete design guidelines. In general, the development of on-chip protection structures is a lengthy expensive iterative process that involves tester design, fabrication, testing, and redesign [74, 36]. When the technology is changed, the same process has to be repeated almost entirely. This can be attributed to the lack of efficient CAD tools capable of simulating the device behaviour up to the onset of failure which is a *3D* electrothermal problem. For these reasons, it is important to develop and use an adequate measure of the EOS robustness of integrated circuits in order to address the on-chip EOS protection issue. Fundamental understanding of the physical phenomena leading to device failures under ESD/EOS events is needed for the development of device models and CAD tools that can efficiently describe the device behaviour up to the onset of thermal failure.

Although several books have been published on the topic, not much in-depth technical discussion dealing with the physical mechanisms of device failures and device modeling for EOS/ESD analysis has been presented. This book is intended for VLSI designers and reliability engineers, particularly those who are working on the development of EOS/ESD analysis tools. CAD engineers working on development of circuit-level and device-level electrothermal simulators would also benefit from the material covered. This book should also be of interest to researchers and first- and second-year graduate students working in semiconductor devices and IC reliability fields.

Book Overview

This book focuses on thermally induced failures of on-chip I/O protection structures. Chapter 1 addresses general aspects of EOS/ESD reliability including an overview of ESD qualification standards and background information on EOS testing and modeling. Chapter 2 presents an overview of common ESD protection practices used in advanced MOS processes. The operation of an nMOS device under EOS/ESD events is qualitatively described along with the mechanisms leading to device failures. The ESD robustness of the nMOS device is also analyzed in terms of process parameters. Chapter 3 presents an analysis of the stress power and the stress current vs. time-to-failure device characteristics as feasible design specifications for EOS protection. The chapter presents an EOS/ESD failure mode distribution across different products, definition of EOS test procedure to reproduce EOS related field failures, and qualification of the power and current device failure characteristics as figures-of-merit to quantify the relative EOS/ESD hardness of semiconductor devices.

In order to solve the heat diffusion equation for thermal analysis, some researchers have resorted to purely numerical methods [62, 67, 61, 48]. Others have combined numerical methods with an analytical solution of linear formulation derived with an appropriate change of variables [92, 7, 79]. However, their work was directed primarily to determining the steady-state temperature profiles across the die under normal operating conditions. Chapter 5 presents a nonlinear thermal simulator iTSIM [22, 30], which was developed to study EOS/ESD thermal failure in ICs, in particular, the sensitivity of the power profiles to major die and package thermal parameters. A set of special boundary conditions has been developed that provides a mixed *2D-1D* simulation capability in iTSIM for computational efficiency while preserving accuracy. Two-dimensional ($2D$) simulated power profiles for various combinations of thermal parameters of the IC die-package structure are presented and compared with experimental data.

In Chapter 6, the EOS performance of gate-coupled nMOS ESD protection structures [34, 35] is analyzed in terms of the experimentally measured power profiles, failure analysis results, and electrothermal simulations; the structures were fabricated in an LDD nonsilicided 0.6 μm MOS process. Device-level eletrothermal simulation was done using the device simulator, TMA-MEDICI [97], which incorporates a lattice temperature model and therefore is capable of simulating device behaviour with self-heating effects. Simulation results matching with experimental data provide a quantitative explanation of the observed failure modes and the ESD/EOS performance dependencies on major layout parameters. Simulation results for a layout design modification of these structures for improved ESD/EOS performance [21] indicate that the current vs. time-to-failure relationship should be used as a measure of the device EOS robustness.

Chapter 7 focuses on circuit-level electrothermal simulation. Temperature dependencies of typical SPICE-model parameters are analyzed for diodes, bipolar and MOS transistors. Additionally, the Miller [72] expression for the avalanche multiplication factor is generalized to incorporate the dependency of the impact ionization rates on the lattice temperature [18, 90]. Thermal instability in electrically stressed circuits or devices is shown to be the result of either the thermally induced negative differential resistance (NDR) in resistive regions, or the junction second breakdown. Furthermore, the results indicate that under typical ESD/EOS stress events (transient in nature), the temperature at which thermal instability takes place is dependent on the level of the stress current. Improved versions of the algorithms in [26] are introduced to assure the convergence of Newton's method for simulating avalanche breakdown phenomena in circuit simulators. Then, iETSIM, a new electrothermal circuit simulator based on the modified nodal analysis approach, is introduced. Examples are presented to show iETSIM's ability to simulate the protection circuit behaviour up to the onset of second breakdown. iETSIM can be also applied to analyze temperature-dependent circuit performances in non-I/O circuits.

Finally, the summary and open research issues are presented in Chapter 9. The User's Guides for the *2D/1D* thermal simulator iTSIM and the electrothermal circuit simulator iETSIM are given in Chapters 5 and 8.

Carlos H. Díaz, Sung-Mo (Steve) Kang, and Charvaka Duvvury
Urbana, Illinois, USA
July 10, 1994

ACKNOWLEDGEMENTS

We appreciate the support and encouragements received from the Reliability Organization of Texas Instruments Incorporated, particularly from Dr. P. B. Ghate and Dr. Donald Denton, and Professor Timothy N. Trick, Head of Electrical and Computer Engineering Department, University of Illinois at Urbana-Champaign. We also wish to express our gratitude to Dr. Larry Wagner, Mrs. Laura Bellay, Mr. William Edwards, and Mr. Don Morgan at the Device Analysis Laboratory of Texas Instruments in Dallas, for helpful discussions on the failure analysis work.

The Technology Modeling Associates has generously donated the electrothermal device simulator MEDICI which has been instrumental in device-level EOS studies. Without research grants from Texas Instruments Incorporated and the Semiconductor Research Corporation, our research on EOS/ESD, which led to writing this book, would not have been possible. We also thank the ICBD R&D Center, Hewlett-Packard, Palo Alto, for the support received for the completion of the final manuscript. Mr. Sridhar Ramaswamy at the Coordinated Science Laboratory has rendered considerable help in the final editing of this book. We would also like to thank Mrs. Lilian Beck in the Publications Office of the Electrical and Computer Engineering Department, University of Illinois at Urbana-Champaign for her professional editing of this book.

1

ELECTRICAL OVERSTRESS IN INTEGRATED CIRCUITS

Semiconductor devices have a limited ability to sustain electrical overstress (EOS). The device susceptibility to EOS increases as the device is scaled down to submicron feature sizes. At present, EOS is one of the major causes for IC failures [47, 24, 69]. Published reports indicate that nearly 40% of the IC failures can be attributed to EOS events. Hence, it is imperative to account for EOS threats early in the design process. EOS embodies a broad category of electrical threats to semiconductor devices due to electromagnetic pulses (EMP), electrostatic discharge (ESD), system transients, and lightning [75]. EOS-related failures in semiconductor devices can be classified according to their primary failure mechanisms into (i) thermally induced failures, (ii) electromigration, and (iii) electric-field-related failures [70, 75]. In general, thermally induced failures are related to the doping level, junction depth, and device characteristic dimensions whereas electric field induced failures are primarily related to the breakdown of thin oxides in MOS devices.

In this Chapter, an overview of EOS and ESD phenomena affecting the ICs is given. Section 1.1 defines the ESD phenomenon and Section 1.2 reviews the impact of ESD on IC technologies. Protection strategies commonly used for reducing ESD effects on ICs are described in Section 1.3. The ESD and EOS models and qualification procedures available today are presented in Section 1.4. Finally, Section 1.5 closes this chapter by reviewing previous works on device modeling for ESD/EOS-induced failures.

1.1 DEFINITION OF ELECTROSTATIC DISCHARGE PHENOMENA

The electrostatic discharge, commonly known as the ESD phenomenon, occurs whenever a charged object is grounded, which results in the release and equalization of the static charge. These ESD events form a subset of transient phenomena known as electrical overstress or EOS, which includes transients on systems, electromagnetic pulses, and lightning. Usually, we consider ESD a threat to the IC chips, which occurs any time an IC chip is touched by human hands or held with metal tweezers, or contacted with a grounded metal object. The latter usually can occur when the devices slide down plastic tubes in a product test engineering environment and the IC's corner pins come in contact with the grounded metal rails. On the other hand, EOS events are commonly associated with the overvoltage and transient spikes in either IC test conditions or during the IC application such as in system boards. The main concern for the microelectronic chips is ESD, which is in the submicrosecond time scale, whereas EOS commonly refers to events other than ESD that encompass time scales in the microsecond and millisecond ranges.

Although much of the reliability focus has been on ESD, EOS is now considered to be a major issue and increased attention to this is expected within the next few years [65, 19]. In this book, both the ESD and EOS-related phenomena will be presented. The main emphasis, however, will be on the EOS modeling and reliability issues.

1.2 IMPACT OF ESD ON IC CHIP TECHNOLOGIES

The ESD effects on the ICs can be varied. At the wafer level, they can appear as damage to the gate oxides which can result in yield loss. As an example, yield loss in the center of wafers was reported by IBM due to oxide rupture [51]. The cause was traced to static charge build-up during jet cleaning which causes oxide breakdown. It was found that this effect can be reduced by an optimization of process parameters such as water pressure and wafer rotation speed. In other cases, yield loss was also traced to a lack of proper protection circuits or poor handling and shipping procedures [62].

At the chip level, the damage due to ESD stress can cause increased leakage at the I/O pins, increased standby current, or, in extreme cases, full circuit failure. Of course, the failures can occur at the board and system levels as well. The IC pin failures can range from very small melt filaments at the junctions to gross damage at the pin site.

ESD/EOS physical effects on ICs can be categorized as thermally induced or field induced failures. Drain junction damage with melted filaments, poly gate filaments, contact metal burnout, or fused metallization are among the thermally induced failure mechanisms associated with ESD/EOS events. On the other hand, dielectric breakdown (gate-oxide rupture) and latent hot-carrier damage are typical field induced ESD-related failure mechanisms. Usually, the damage signature indicates whether it was ESD-related or EOS-related. In general, ESD-related damage is associated with small failure sites, e.g. oxide pin-holes, polysilicon filaments, or junction deformation preferentially at diffusion corners. On the other hand, EOS damage can be quite drastic and cover most of the area around the pin. This damage is commonly caused by long lasting ($> 1\ \mu s$) conditions for which the device failure (generally associated with thermal runway) happens early during the stress event. In this way, what was a probably small failure site just after the onset of thermal runway, will invariably grow in size as a result of the additional energy deliver to the structure for the rest of the stress event. Regardless of the stress event, ESD or EOS, the failure sites can be traced to the protection circuits in general. However, the protections devices or the coordination among the protection structures within an IC may prove to be ineffective for certain types of ESD/EOS events, case in which failure sites will also be found in the internal circuitry. CMOS latchup is also due to overvoltage and the resulting damage can be potentially confused with EOS. Thus, EOS failure analysis often requires a careful evaluation of the cause and the effect. The details of the EOS failure mechanisms are discussed in this book.

Besides showing the obvious damage effects, ESD can also be subtle. For example, the ESD-related damage can cause an oxide or a transistor to degrade gradually with time during the circuit operation and lead to an eventual chip failure. These are 'latent' or hidden failures [66]. Not much work has been done in this area so far and considerable controversy exists if this indeed is a real threat. Unlike the ESD event, EOS events are not known to cause any latent effects.

As advances are made in the process technologies, the ESD effects seem to be getting worse. This is due mainly to the thinner gate oxides and shallower junctions. The complexity also increased as we moved from NMOS to CMOS and BiCMOS. Also, GaAs devices have been reported to suffer from ESD damage [86]. Of course, for all of these technologies, because of high packing densities, EOS failures can also be expected to be higher. For any technology, a reliable protection strategy is needed along with in-depth understanding of the failure phenomena. The work presented in this book offers a systematic experimental and modeling approach to this end.

1.3 PROTECTION STRATEGIES FOR REDUCING ESD EFFECTS

Any EOS or ESD event can be catastrophic to the IC if proper protection methods are not implemented. These methods fall into external and internal chip precautions. The external methods for ESD are achieved by storing them in conductive containers which form a Faraday shield and maintain static-free work environments where the chips are handled. For EOS, proper grounding of the equipment and shielding can minimize the impact.

The internal methods require the placement of adequate on-chip protection devices in the I/O frame and in the power supply pins [40, 33] to absorb the ESD energy. The on-chip protection scheme should be such that there is an explicitly robust path for the ESD currents to flow between any pair of pins. In general, pad protection networks shunt I/O pins to the ground bus under stress events. For each input pin, a dedicated protection network that is completely passive under normal operating conditions has to be added. For each output pin, the ESD protection level is determined by the intrinsic robustness of the output buffer transistors plus that of the deidcated protection device(s).

The most challenging aspect for ESD/EOS reliability is the protection circuit efficiency, especially in light of the highly susceptible advanced CMOS technologies. Higher current densities and reduced ballasting resistances are known to favor localization of stress currents and to degrade the electrostatic discharge performance of I/O protection devices in advanced MOS processes with shallow LDD junctions and silicided diffusions [16, 38, 37]. The Human-Body-Model Electrostatic Discharge (HBM-ESD) is the method most widely used to assess the effectiveness of the on-chip ESD protection structures [71]. Effective protection circuits can now withstand more than 8 kV HBM-ESD [13, 34, 35]. However, HBM-ESD events represent only a small portion of the whole EOS spectrum. Characteristic failure times for EOS events might be as high as several tenths of a millisecond compared to the HBM-ESD submicrosecond time constant. Therefore, even with the best on-chip ESD protection structures available, the overall EOS protection is not necessarily adequate. In other words, meeting stringent ESD specifications does not warrant EOS protection.

The chances for optimum protection design will improve if both the device self-heating effects and the interaction of the different circuit elements during electrical stress can be accurately modeled. In this book, the main theme is to illustrate the use of a modeling approach for understanding and designing effective protection devices for ESD and the more general EOS events. The approach given here is only for NMOS, CMOS, and BiCMOS technologies. Some EOS failures for the bipolar technologies are also discussed.

1.4 ESD MODELS AND QUALIFICATION

Currently, most commonly used models to describe various categories of ESD pulses affecting ICs during handling are [71, 55, 81, 82]

- Human Body Model (HBM),
- Machine Model (MM), and
- Charge Device Model (CDM).

The basic model for ESD protection is the Human Body Model (HBM) intended to represent the ESD caused by human handling of ICs. The model describes the discharge of the body capacitance (about 100 pF) when a finger touches the device pin. If the initial human body charge build-up is to the level of 1000 V or more, the stress current, which is limited by the body resistance (about 1500 Ω), is high enough to destroy the device connected to the pin. The HBM equivalent circuit is shown in Fig. 1.1(a). The 100 pF capacitor is charged with a high voltage supply and then discharged through a 1.5 $k\Omega$ resistor onto the pin under test. The HBM-ESD is the most widely used method to qualify the ESD performance of on-chip protection circuits and is standardized according to [71]. Typically, HBM events occur at 2-4 kV in the field, hence, protection levels of this range are necessary.

Besides human handling, contact with machines is also an ESD-type stress event. Since the body resistance is not involved here, the stress is severe at relatively higher current levels; thus, protection levels of 200 V for this model usually ensure device reliability. The MM-ESD is intended to model the ESD produced by a charged object making contact with ICs during device bonding, assembly, or testing. MM-ESD testers deliver damped oscillating ($\sim 20\ MHz$) stress currents (first peak value in the order of 1-10 A) to the device under test. A schematic representation of an MM-ESD stress tester is shown in Fig. 1.1(b). In contrast to the HBM-ESD test method, there is no unique definition of the MM. Two generally used MM specifications originated from Japan and the Phillips Research Laboratories [82].

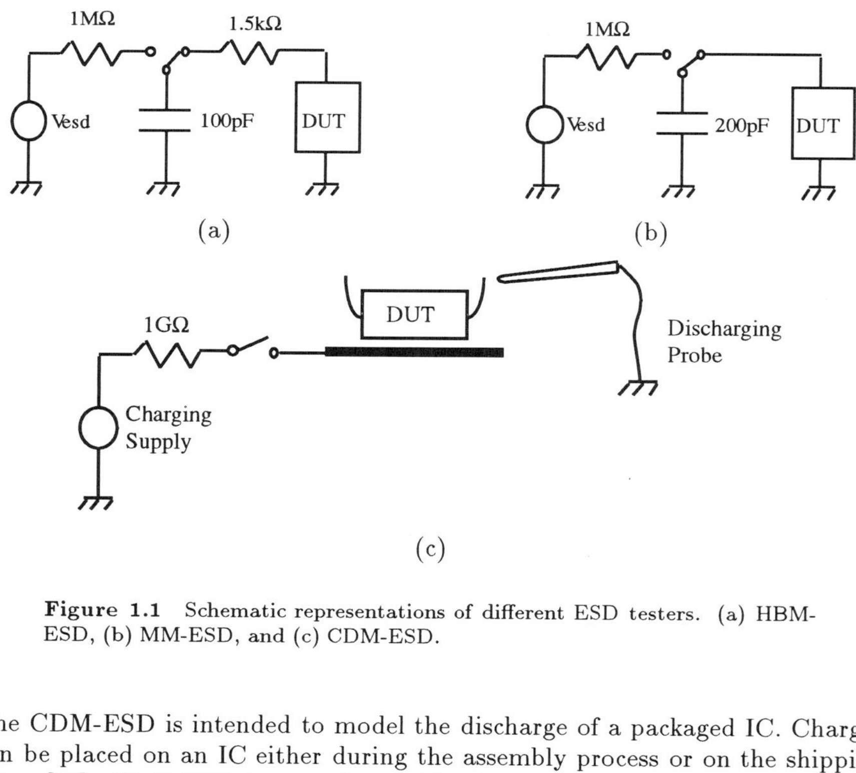

Figure 1.1 Schematic representations of different ESD testers. (a) HBM-ESD, (b) MM-ESD, and (c) CDM-ESD.

The CDM-ESD is intended to model the discharge of a packaged IC. Charges can be placed on an IC either during the assembly process or on the shipping tubes [81]. CDM-ESD testers electrically charge the device under test (DUT) and then discharge it to ground, thus providing a high-current short-duration ($\sim 5\ ns$) pulse to the device under test. As in the MM-ESD case, there is no industry agreement on the CDM-ESD test method specifications. One well-known CDM-ESD specification is from AT&T Microelectronics [55, 81]. A schematic of the CDM-ESD tester proposed in [81] is shown in Fig. 1.1(c).

It should be recognized that it is practically impossible to fabricate ESD testers that implement the theoretical HBM and MM waveform specifications. The

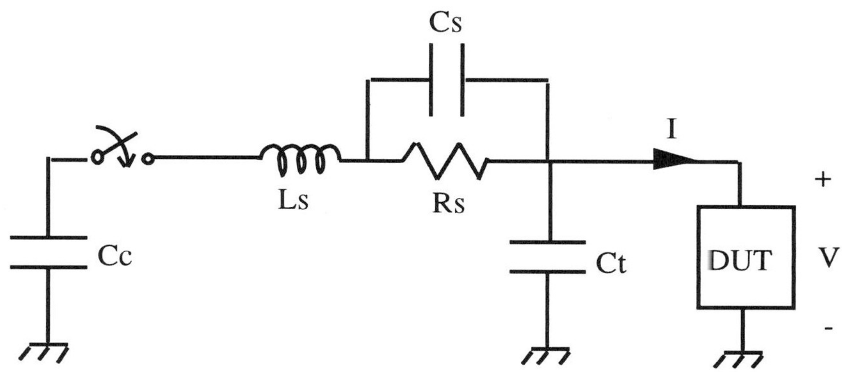

Figure 1.2 Simplified lumped-element model of HBM-ESD and MM-ESD testers.

lumped element model shown in Fig. 1.2 has been proposed to model actual HBM-ESD and MM-ESD simulators [14, 82]; typical values of the model elements are given in Table 1.4 according to [82].

Table 1.1 Component and typical parasitic-element values for the lumped-element model of ESD testers.

Component	HBM	MM
Cc $[pF]$	100	200
Rs $[\Omega]$	1500	25
Ls $[\mu H]$	5	2.5
Cs $[pF]$	1	0
Ct $[pF]$	10	10

A good protection device should be able to safeguard ICs from any of the above stress model tests.

1.5 EOS MODELS AND QUALIFICATION

The EOS test is more complicated due to the wide spectrum of electrical characteristics of stress events to which an IC may be subjected in its lifetime [75]. Currently, there exist no EOS standards or quantitative EOS design objectives to achieve, thus limiting and/or delaying the designer's attention to the EOS problem. Constant power pulses are commonly used for EOS testing [75, 76, 42, 23]. Such EOS stressing is easy to generate consistently and is also amenable to simple analysis. For thermally induced failures, failure thresholds for any arbitrary stress waveform can, in principle, be obtained from the failure threshold derived under pulse-stress conditions and given in terms of the power-to-failure vs. time-to-failure relationship $P_o(t_f)$. This is based on the *waveform conversion technique* known as the Duhamel formula [43, 75]

$$1 = \int_0^{t_f} P(\tau) \frac{d}{d(t_f - \tau)} \left[\frac{1}{P_o(t_f - \tau)} \right] d\tau \tag{1.1}$$

This formula can be derived from purely thermal considerations that involve the solution of the time-dependent heat diffusion equation and the specification of a critical (failure) temperature. For these reasons, square pulse testing has been used in most of the EOS published work [100, 96, 75, 73, 76, 23]. Constant current pulses lasting 100 - 250 ns are also being used in device characterization to extract ESD-related parameters [5, 77, 75, 10]. High power pulse generators can be used as EOS stress sources for events lasting longer than 1 μs, while charged transmission lines [63] are commonly used for submicrosecond stressing of devices.

1.6 PREVIOUS WORK ON ESD/EOS DEVICE FAILURE MODELING

Previous models of ESD/EOS device failures can be classified as either thermal or electrothermal models according to the extent to which the electrical description enters in the particular model.

Thermal models focus on the thermal aspect of the problem and limit the electrical description to the calculation of the power delivered to the device by assuming temperature-independent electrical parameters. In this case, it is assumed that the onset of failure can be defined in terms of a critical temperature (failure condition) for each device. The heat diffusion equation is solved in a closed form to find the temperature distribution as a function of power generation and time. Failure is declared when one point in the structure reaches the specified critical temperature. In the late 1960s, Wunsch and Bell [100] demonstrated that the experimental data relating the power required to cause device failure to the time-to-failure could in part be explained by solving the linear *1D* heat diffusion equation. The work has been revised since then [96, 7]. A more extensive theoretical treatment for thermal failure of semiconductor devices was published in 1990 [43]. There, the *3D* time-dependent heat diffusion equation is solved for an infinite medium with constant parameters. The pulsed heat source is a localized parallelepiped whose dimensions have to be determined experimentally. Upon specifying a critical temperature at which failure is declared, it is possible to establish (i) the waveform conversion equation and (ii) a set of closed-form expressions for asymptotic behaviour of the relationship between the power and the time-to-failure. Figure 1.3 shows the expected asymptotic behaviour. In principle, the model parameters *a, b, and c* could be estimated based on device geometry and major process parameters. However, the parameter values determined in this fashion will in general require additional tuning before the model can closely represent the experimental data [43, 5, 4, 27, 28]. This is due to the fact that the model does not account for the presence of the passivation layer, such as SiO_2, and for the nonlinear behaviour of the thermal parameters of silicon [90].

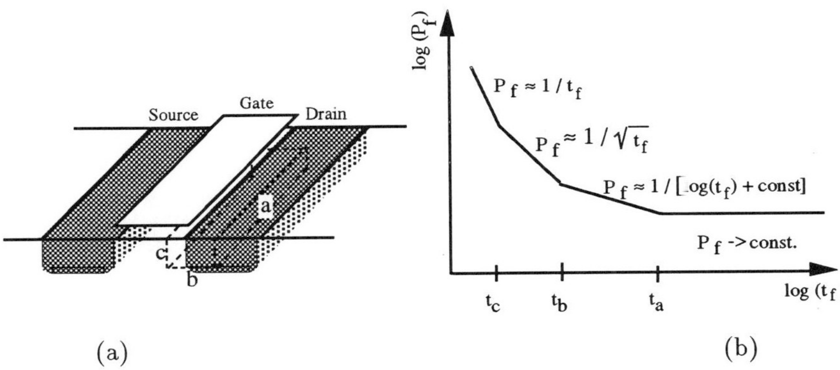

Figure 1.3 (a) The definition of the heat source for the thermal failure model applied to an MOS device whose drain is under positive stress with respect to source. (b) Analytical thermal failure model predictions of the asymptotic behaviour for the power-to-failure relationship.

Electrothermal models are expected to describe the device behaviour up to the onset of thermal breakdown (cf. [87, 94, 50, 73, 5, 4]) which causes current localization and eventual device failure (thermal runaway). Thermal runaway in a circuit component is a positive feedback process produced by the coupling of the device self-heating and the temperature-dependent electrical parameters under stress conditions. The second breakdown is a synonym for the thermal breakdown in semiconductor junctions operated in reverse mode.

The most widely used electrothermal numerical models solve the coupled set of semiconductor device equations (drift-diffusion model) and the nonlinear heat diffusion equation [90, 99]. The coupling is provided by the lattice temperature dependency of the semiconductor electrical parameters and by the heat generation term. Although the basic set of equations governing nonisothermal behaviour in semiconductor devices was known by 1950 [90], in 1978 Adler [2] proposed a more physically plausible, but still heuristically motivated, form for the heat generation term. Adler's work was related to steady-state power dissipation in thyristors. Recently, Wachutka [99] has given a more rigorous treatment of the heat generation and conduction terms for device electrothermal modeling by using the laws of phenomenological irreversible thermodynamics. Currently, there are various CAD tools capable of performing this kind of device-level fully coupled electrothermal simulation [97, 11, 64, 56].

2

NMOS ESD PROTECTION DEVICES AND PROCESS RELATED ISSUES

In this chapter, the ESD resistance of a given MOS technology is analyzed in terms of the ESD performance of the most commonly used protection structure, namely the nMOS device itself. Section 2.1 gives an overview of ESD physical phenomena in the nMOS device and the design criteria for ESD protection devices. Section 2.2 reviews the failure modes in nMOS ESD protection devices. Sections 2.3 and 2.4 describe ESD nMOS protection schemes and the impact that advanced processing techniques such as LDD and silicidation have on the nMOS ESD robustness. Finally, Section 2.5 discusses protection concepts for advanced MOS technologies.

2.1 ESD PHENOMENA IN NMOS DEVICES

The most commonly investigated and well-understood phenomenon is the ESD behavior of an nMOS transistor. Depending on how they are incorporated, these transistors can either be most susceptible to ESD damage or actually form robust self-protection devices. To illustrate this point, we consider the operation of an nMOS device under high current conditions. The basic I-V characteristics are shown in Fig. 2.1 where the gate and the substrate are tied to ground. Here V_{t1} is the collector-base breakdown voltage of the *npn* with the emitter open-circuit BV_{cbo}. During the ESD event, the device operates primarily in the snapback mode where V_{sp} is the snapback voltage. At higher stress levels, the device could go into second breakdown which usually results in device failure due to thermal effects. This is discussed in more detail in the next section. Presently, a typical nMOS device will clamp the ESD voltage at

V_{t1}, which is generally lower than the thin gate oxide breakdown of the technology. For example, for 1 μm technology, the breakdown voltage can be around 14-15 V with a snapback holding voltage of 8-9 V. The gate oxide for these same technologies is typically 200 $\AA$ with a dc breakdown voltage in the 20 V range. Thus, the clamping of this nMOS device can easily protect the gate oxide of an input buffer. Note, however, that with the advent of the submicron technologies, much thinner gate oxides (65-80 $\AA$) will be used. A condition in which the oxide breakdown voltage is lower than the drain avalanche can occur making this protection concept ineffective. For these conditions, techniques to lower the breakdown of the drain junction will help. In addition to protecting the gate oxide, the nMOS device can function as a self-protection device. Usually, in VLSI chips, the large nMOS output buffer transistor is used as the protection device. That is, the parasitic *npn* device handles the high ESD current discharge. To understand more about the design and limited effectiveness of this device, the common ESD failure modes have to be considered.

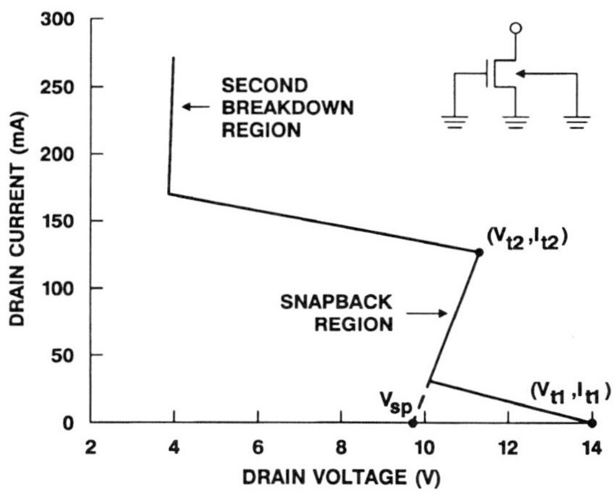

Figure 2.1 The breakdown *I-V* characteristics of an nMOS transistor with gate and substrate grounded.

2.2 FAILURE MODES IN NMOS

As mentioned before, the most common failure modes are thermal in nature, that is, due mainly to the heat generated with the device under stress. In an nMOSFET, the heat generated at the reverse-biased drain junction during an

ESD pulse can lead to the failure modes shown in Fig. 2.2. For the conventional abrupt junction processes, when the drain junction is stressed with the source and substrate grounded (common), the breakdown will occur at the cylindrical portion of the drain junction causing maximum heat generation at this location. This heat can spread either to cause a metal melt filament from source to drain by elevating the temperature near the contact metallization, or result in polysilicon melt filaments by spreading to the gate directly above the thin oxide. The first mode occurs due to what is known as 'electrothermomigration'. Moving the drain contact farther away from the gate edge can eliminate this mode [85]. The second mode occurs due to the isotropic nature of the heat source which can spread through the thin insulating gate dielectric and melt the polysilicon. Since the gate is encapsulated, the heat cannot escape, resulting in the gate-to-drain melt filament. Once this is formed, the electric field shifts and another filament is formed on the source side. This type of polysilicon filament is illustrated for a thin oxide nMOS in Fig. 2.3. A thick field oxide device will not show this type of failure mode because of the relatively higher thermal insulation. Even in a thin oxide nMOS device, this failure mode can be avoided by using layout techniques that can provide more uniform heating throughout the device. These are discussed in the next subsection as well as in Chapter 6.

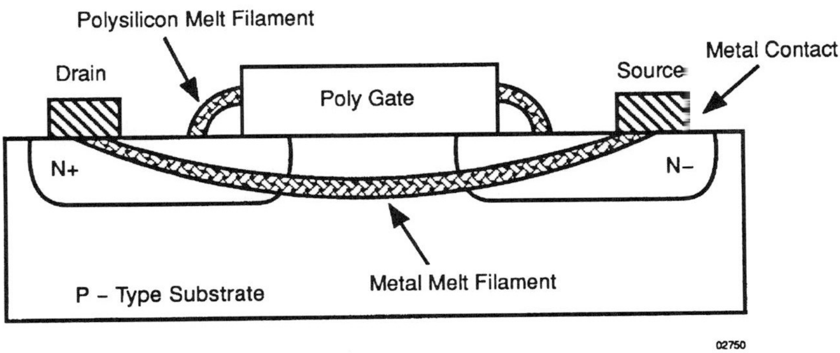

Figure 2.2 Typical failure modes caused by device self-heating under ESD stress in an nMOS transistor.

Referring back to Fig 2.1, at higher current levels, the low on-resistance of the *npn* results in a relatively small voltage increase until a second breakdown at V_{t2} and I_{t2} is reached. This condition occurs at a high temperature when the local thermal carrier generation is high enough to sustain most of the stress current as explained in Section 6.1. Then the current in the surrounding higher

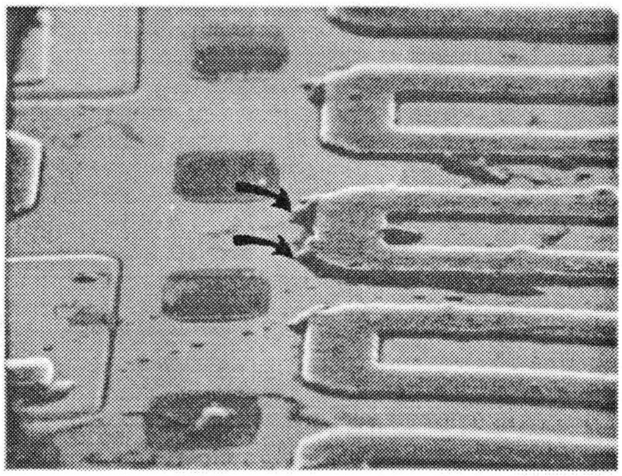

Figure 2.3 SEM photograph of an ESD failure site in an nMOS transistor.

resistance regions is shunted through the low resistance region of the local area. Thus, the second breakdown condition usually leads to filament formation at the drain junction. If the ESD protection devices are operated only in the snapback region, no leakage current increase would occur after an ESD event. The filament formation, the cause for ESD failures, can be inhibited by moving the drain contact away from the gate edge, thus providing the needed ballasting resistance.

In addition to the filaments at the drain, contact spiking can occur under second breakdown as the current density is increased. That is, the joule heating in the contacts can lead to spiking. The contact spiking is reduced in technologies that use barrier metals (such as TiW) in the contacts.

2.3 PROTECTION TECHNIQUE USING NMOS DEVICE STRUCTURES

The effective ESD protection level realized by an nMOS device essentially depends on three factors: the process technology, the device design, and the device layout. The effects of process technology are discussed in the next subsection while the latter two effects are discussed here.

The protection level offered by an nMOS device is usually measured in V/μm.

If a device offers 10 V/μm, then only a 200 μm wide device is needed to meet the often required minimum 2 kV of protection for the Human Body Model. However, it is not easy to achieve the required protection level because it does not scale uniformly with device width. Nonetheless, proper layout techniques can help in this regard.

As mentioned earlier, the drain contact-to-gate spacing is a critical parameter and should be maximized to improve the ESD level offered by an nMOS device. Furthermore, a more uniform type of layout in Fig. 2.4, called the 'ladder structure' is needed. This type of layout is used to achieve uniform current density by using multifingers for a large device. Here, the parasitic resistance R_F is composed of the metal finger resistance, contact resistance, and the diffusion sheet resistance. The parasitic sheet resistance, R_S, consists of the ground bus resistance that connects the device source fingers. A minimum ratio for R_S/R_F should be maintained to obtain the best possible performance from this device [41]. Note that while maximum R_F is desirable, contact resistance itself should not be high since excessive contact heating can cause damages. The ballasting resistance with the increased drain-to-gate spacing will thus contribute to R_F and help in the uniform turn-on of the device. A typical value for this optimum spacing is 6 μm. Increasing the spacing beyond this value usually does not result in a further increase in the protection level from the nMOS fingers, because the failure is eventually determined by the heating at the drain junction. The uniformity of the device can also be obtained by decreasing R_S, which can be achieved by making the ground bus as wide as possible. Typically about 15 μm is sufficient. The device channel length is the other remaining design parameter. It has only a minor impact. Nevertheless, a minimum channel length is preferred since this results in smaller V_{sp} in Fig. 2.1, and reduces the power dissipation during ESD.

Normally, large output transistors of the output buffer, both pull-up to Vdd and pull-down to GND, can operate as protection devices provided the layout implements the ladder structure with an optimum drain contact-to-gate spacing. In CMOS circuits, the p-channel pull-up device can play an important role. As shown in Fig 2.5, in a CMOS buffer, the p-channel pull-up device provides a diode to Vdd for positive stress pulses. But even then, after the Vdd capacitance is charged up, the voltage build up at the pad will take the pull-down n-channel into reverse breakdown. As a result, failure sites after ESD stress are almost always observed only on the n-channel transistor to GND. Therefore, for good ESD protection, the n-channel devices in CMOS output circuits should be designed to be robust. As the fabrication process progresses toward deep submicron feature sizes, the n-channel device has shown degradation to ESD. These issues are briefly discussed in the next subsection.

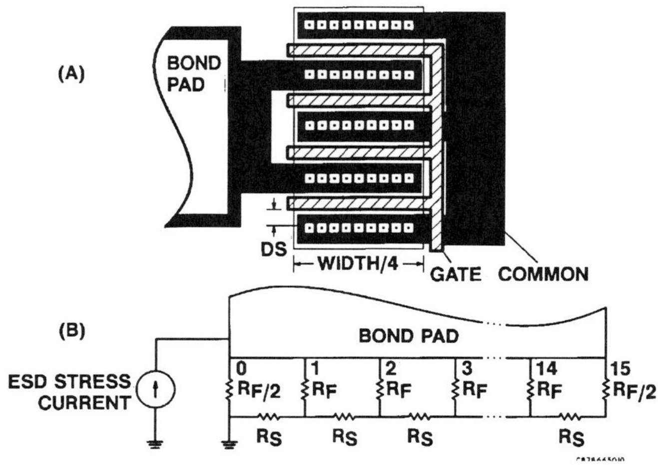

Figure 2.4 (a) Ladder structure layout for nMOS transistor and (b) its circuit representation.

2.4 THE IMPACT OF PROCESS TECHNOLOGIES ON NMOS ESD BEHAVIOR

As the IC chip designs progressed from 2 μm to 1 μm and to submicron channel lengths, various advances have been made in the corresponding process technologies to improve the circuit performance as well as to counter other reliability threats. For example, as the transistor feature sizes are scaled down, the increase in the peak electric field at the drain junction causes concerns for the hot carrier reliability. These peak fields cause a higher rate of impact ionization that can result in both interface trap generation and charge trapping in the oxide. Such effects degrade the transistor and subsequently the circuit performance. In the early 1980s, graded junctions and lightly doped drain (LDD) structures for transistor source and drain became popular, mainly to improve the device resistance to hot carrier degradation. Figure 2.6 shows the junction grading with double or single diffused LDD structures in comparison with the abrupt junction. This junction grading is done to reduce the peak electric field at the drain and improve the hot carrier reliability. The same reduced electric field was expected to decrease the Joule Energy (J.E.) heating at the junction under ESD and improve the ESD reliability as well. In a surprising twist, however, the ESD levels actually degraded for the LDD transistors [91, 68]. There

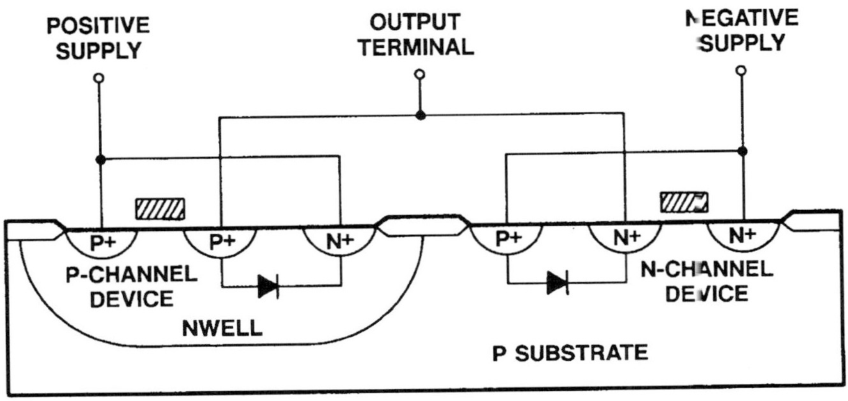

Figure 2.5 Cross-section of a CMOS output buffer.

are two reasons for this. First, the LDD devices, because of lower substrate current (due to the lower peak field), show increased V_{sp}, which increases the power dissipation during ESD as the device clamps at this voltage during most of the ESD event. Experimental works have shown that indeed there is some correlation to the V_{sp} value and the maximum ESD level [15, 39]. However, this effect is somewhat weak and does not fully account for the ESD degradation of LDD devices. Second, it is believed that the other reason is the relatively shallower junction of the LDD which increases the current density, and, even with the reduced electric field, actually results in a higher J.E. Typically, the ESD performance level drops from about 12-15 $V/\mu m$ for an abrupt junction device to about 8-10 $V/\mu m$ for the LDD devices. These numbers will vary considerably depending on the particular process technology. In fact, drain engineering has been reported [3, 20] to counter the LDD effect. In summary, the LDD devices do degrade the ESD levels but a complete understanding of this phenomenon has not yet been obtained. There have been some electrothermal simulation studies in an attempt to understand this further [4].

Besides the LDD, silicided diffusions have also been introduced for the source and drain of the advanced transistors as also shown in Fig 2.6. The decreased sheet resistance with the silicides has a positive impact on circuit speed. However, the silicide effect on ESD has been disastrous [68, 38]. The nMOS ESD levels dropped to as low as 5 V/μm with the silicided diffusions. The main reason for this is that the clad silicides virtually eliminate the desired ballasting resistance between the drain contact and the gate edge. Consider the

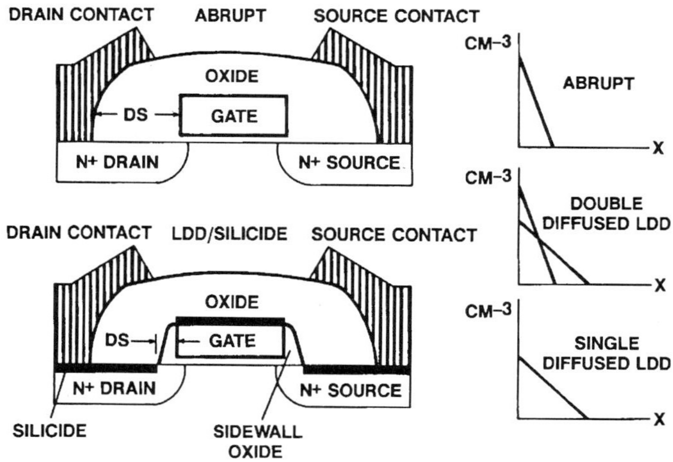

Figure 2.6 Cross-sections of conventional abrupt junction LDD junction transistors. The clad silicide of drain and source regions is shown for the bottom device.

cross-section shown in Fig 2.6. The only obvious way to improve this ballast resistance is to increase the sidewall thickness which separates the gate edge from the silicide contact. But this is practically limited to only about 0.5 μm. The clad diffusions actually modify the source/drain resistance due to the effect of the silicide/silicon interface and result in nonuniform current density [89]. Figure 2.7 shows the peak current density at the drain and source sidewall edges [83, 32]. The peak field at both edges leads to symmetrical formation of melt filaments at drain and source. This is shown in the SEM photograph in Fig. 2.8. To overcome the silicide effect, process changes have been tried but with only a minimal effect. An obvious an effective solution is to block the silicide formation, but is expensive because of the required additional mask level.

Other process advances, such as smaller contacts, thinner metal lines, or more highly doped substrates, have been observed to have no apparent effect on the ESD levels. Since LDD and silicides are the major effects, improvements to the protection circuit design must be made to overcome the problems or at least optimize the protection devices to give the maximum level of performance possible for a given technology. Some of the issues are discussed in the next subsection.

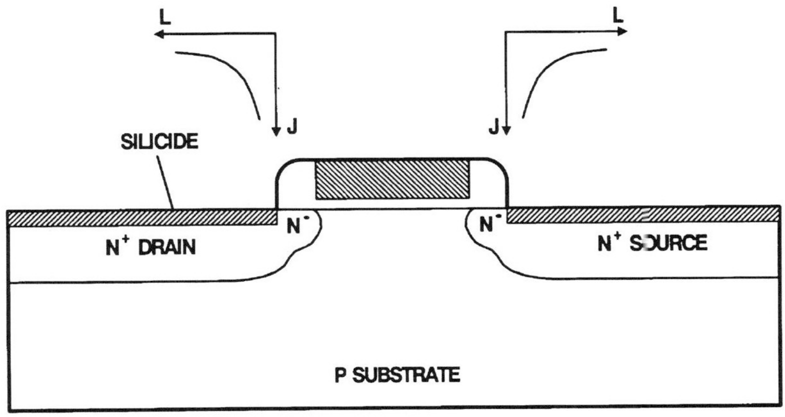

Figure 2.7 Current density distribution at the drain and source edges of a silicided nMOS transistor.

2.5 ADVANCE NMOS DEVICE PROTECTION CONCEPTS

As discussed so far, the basic nMOS device can be designed to give robust ESD performance, although the advances in fabrication processes tend to degrade the protection levels. Besides the process technologies, the chip function can also have an effect on the protection level. For DRAM chips, the floating substrate can improve the turn-on efficiency for a large multifinger nMOS device, typically used for output buffer design. During an ESD event in DRAM chips, a small amount of substrate current is enough to raise the substrate potential and turn on all the lateral bipolar devices for uniform conduction. Unfortunately, this is difficult to achieve for the common grounded substrate logic chips and the nMOS device ESD levels can become unpredictable [77]. This random behaviour is illustrated in Fig. 2.9 where the ESD failure distribution for a 20-finger grounded gate nMOS is shown with a sample size of 28 data points. It is clear that the failure levels range from below 2000 V to above 7500 V depending on the number of fingers turned on, as indicated in the figure. The reason for this is quite simple. As each finger turns on, the voltage has to build up to turn on the next subsequent *npn*. With the substrate being held close to the source potential, usually the failure current level is reached only after two or three fingers turn on. Note that in some cases more fingers did turn on

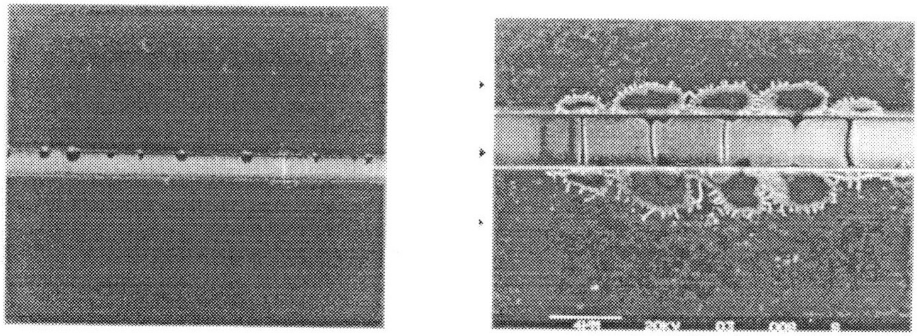

Figure 2.8 Symmetrical drain-to-source filaments (under the gate) caused by ESD damage to a field-oxide transistor fabricated in a silicided process.

to give the higher levels. It is speculated that after one or two fingers go into snapback enough trigger current is supplied to turn on the subsequent fingers. But as noted, this is a random event and cannot be expected to give full ESD reliability.

A new approach to improve the uniformity in a multifinger device is to provide a coupling for the nMOS gate to turn high during ESD [34]. This device concept, called the gate-coupled nMOS (GCNMOS), is shown for a cross-section in Fig. 2.10. Here the nMOS gate is tied to GND through a field oxide device while the field oxide device (FOD) gate is controlled by the pad tied to the drain of the nMOS. Now the ESD pulse at the nMOS gate, due to the presence of ESD on the pad tied to drain and the gate-drain overlap capacitance, will momentarily pull the gate above the transistor threshold voltage level.

SPICE simulations of this effect are shown in Fig. 2.11. The nMOS gate transient briefly turns on the nMOS device and causes impact ionization to create electron-hole pairs. The holes collected by the substrate would elevate the local substrate potential to facilitate the uniform turn-on of all fingers. The effectiveness of this device performance will depend on how well the gate transient is controlled. A different view of this phenomenon is that with the gate transient above V_t, a significant amount of substrate current is generated. This would lower the bipolar trigger voltage V_{t1} of Fig. 2.1 closer to V_{t2}. This effect is shown in Fig. 2.12 where the measured breakdown voltage is plotted as a function of gate bias for different process technologies. Note that for most cases a gate bias of about 1 V is enough to minimize V_{t1}. Thus the turn-on of each finger will require only a smaller voltage excursion. This is illustrated in the

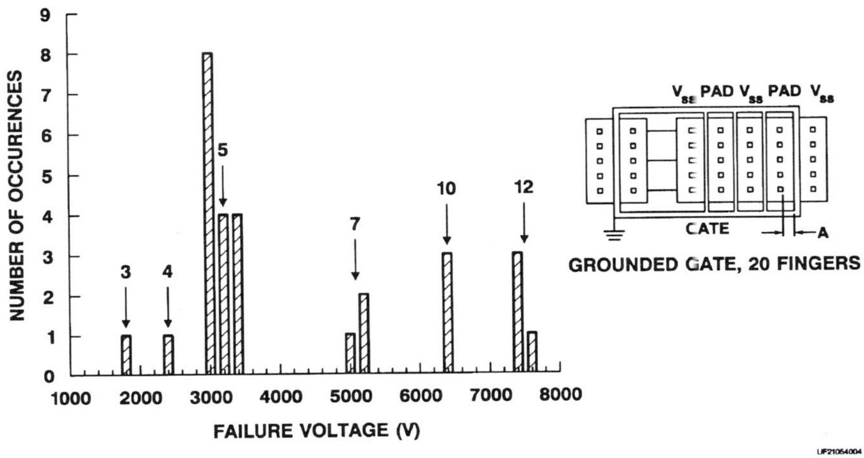

Figure 2.9 Random ESD failure distribution for a grounded multifinger gate nMOS transistor.

I-V curve of Fig. 2.13. Notice that in contrast to the grounded gate device, the multifinger turn-on is observed with local voltage maxima.

The improvement of the ESD failure thresholds with this concept is illustrated in Fig. 2.14(a) for an LDD process with no silicide and in Fig. 2.14(b) for an LDD process with silicided diffusions. It is obvious that not only did the performance improve dramatically, but also an almost ideal width dependency was obtained reflecting the uniformity in device conduction. Gate-coupled devices can be expected to maintain uniform conduction behaviour for the longer EOS kind of stress events. The GCnMOS device will be extensively discussed in Section 3.3 and Chapters 6 and 7 for both ESD and EOS applications using experimental data and electrothermal simulation tools.

In addition to the GCnMOS device, there are other device techniques to improve the ESD levels in advanced technologies. For example, lateral SCR or *pnpn* devices that clamp with a holding voltage below Vdd can be used to protect the output transistors [84, 13]. A cross-section of the basic lateral SCR device is shown in Fig. 2.15(a). This device has a trigger voltage on the order of 40 V corresponding to the n-well to substrate breakdown. The SCR trigger voltage can be substantially reduced if an $n+$ tap is added at the edge of the n-well as shown in Fig. 2.15(b). In this figure, a polysilicon gate has also been added to minimize the snapback voltage of the lateral npn that is present in

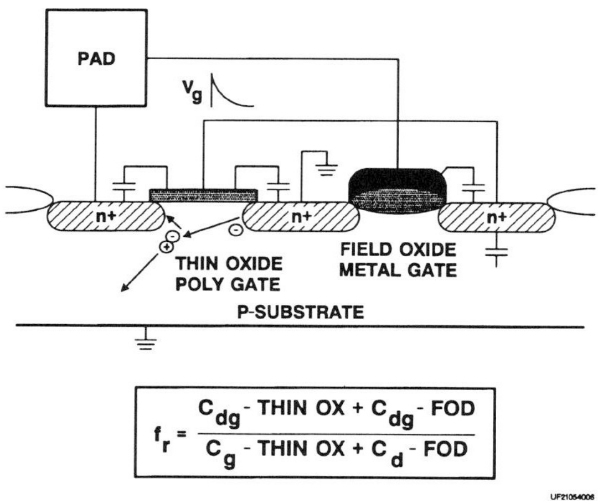

Figure 2.10 Cross-section of a gate-coupled nMOS (GCnMOS) transistor.

the anode just before the regenerative SCR action is triggered [13]. The SCR devices are now widely used for input protection and in some cases for output protection [36, 9]. For outputs, this protection technique requires an isolation resistor between the output device and the SCR so that when the nMOS goes into breakdown and the parasitic *npn* transistor conducts, the voltage build-up due to the IR drop across the resistor will elevate the pad voltage and trigger the SCR connected to it. This is necessary because the SCR trigger voltages are typically higher than the nMOS breakdown voltage. Although this approach is very effective, it has a few drawbacks. First, the series resistor will degrade the output buffer performance. Second, for technologies with thinner epitaxial substrates (to prevent CMOS latchup), the SCR becomes less efficient in its turn-on. Recently, a bi-modal triggering scheme for the LVSCR has been proposed [31]. In this approach, an active circuit is used to drive the gate of the LVSCR in such a way that under ESD/EOS events, the trigger voltage is much lower than the corresponding nMOS breakdown voltage thus eliminating the need for a series resistance in the output buffer. During normal operation, the active circuit effectively grounds the LVSCR gate thus restoring the LVSCR trigger voltage to a value close to the nMOS breakdown voltage; the LVSCR becomes passive during normal operation.

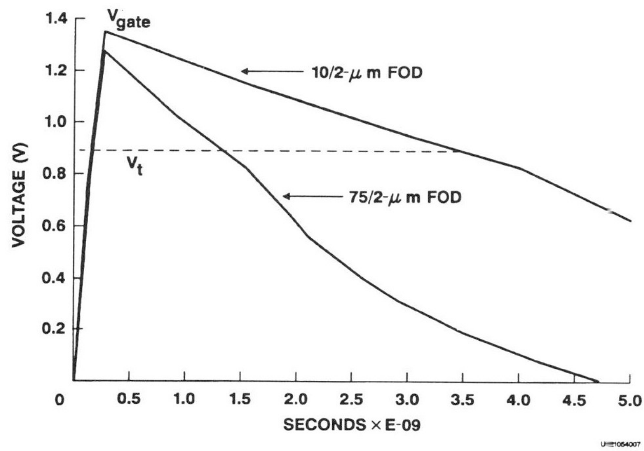

Figure 2.11 SPICE simulations of gate coupling for two different sizes of the field oxide device (FOD).

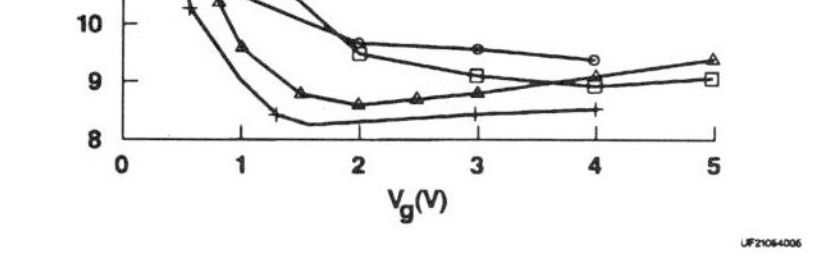

Figure 2.12 Measured nMOS transistor breakdown as a function of gate bias for different CMOS technologies. The y-axis is the value of V_{t1} as defined in Fig. 2.1.

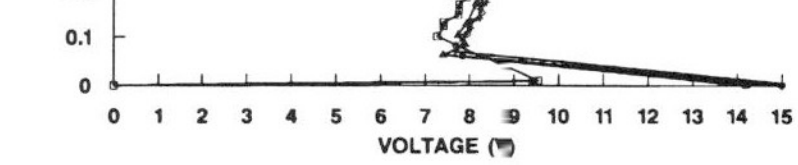

Figure 2.13 Measured $I - V$ breakdown characteristics of a grounded gate nMOS transistor compared to a gate-coupled nMOS device.

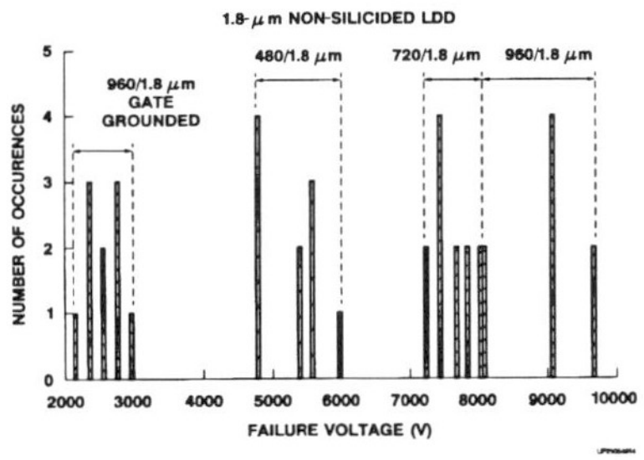

(a)

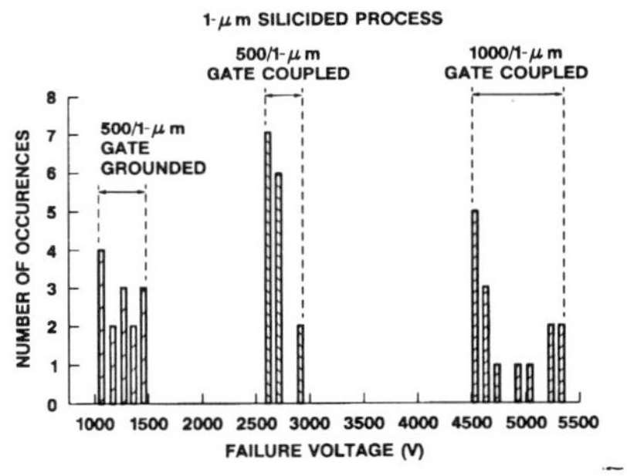

(b)

Figure 2.14 ESD failure distributions for GCnMOS transistors fabricated in (a) a nonsilicided LDD process and (b) a silicided LDD process. Different device widths are indicated in the figure. A grounded gate case is also shown for comparison in both cases.

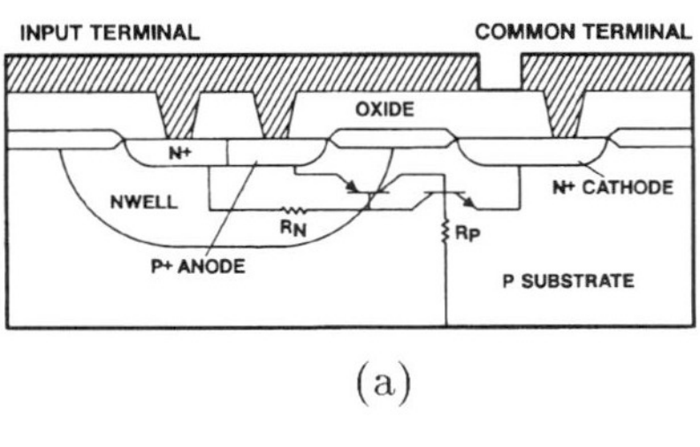

(a)

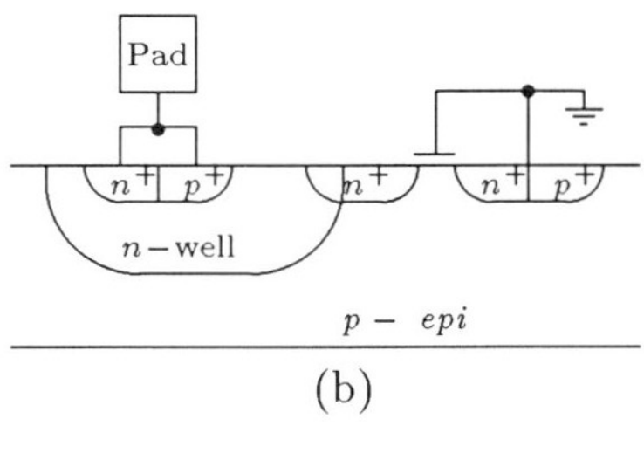

(b)

Figure 2.15 (a) Cross-section of a lateral SCR. (b) Cross-section of a grounded-gate low trigger voltage SCR (LVSCR).

3

MEASURING EOS ROBUSTNESS IN INTEGRATED CIRCUITS

In this chapter, the power vs. time-to-failure (power profile) and the stress current vs. time-to-failure (current profile) device characteristics are proposed as feasible measures to determine the relative EOS hardness of integrated circuits. Experimentally measured power and current profiles together with failure analysis techniques can be used to pinpoint layout design weaknesses. Bipolar integrated circuits with different I/O ESD protection designs and nMOS gate-coupled protection devices are characterized for HBM-ESD and EOS under unipolar stress conditions. It will be shown that, according to measured power and current profiles, good ESD performance is not a sufficient condition to assure EOS robustness.

To identify EOS/ESD-related dominant failure mechanisms across the various IC product families, a set of EOS/ESD susceptible devices that could be tested in volume (low cost) is defined. A test method is specified to consistently reproduce observed field failures in those devices. Unipolar positive stress waveforms were applied to the input and output protection transistors present in those ICs. One reason for this choice is that under reverse bias conditions the protection transistors have the highest power dissipation (self-heating) conditions that might lead to thermal runaway and device failures. Another reason is that if one intends to unveil and understand the failure process under any bipolar stress event, one first has to know precisely the phenomena leading to device failure for each stress polarity independently. All of the results reported in this chapter are based on unipolar positive stress events applied to (i) the I/O protection transistors found in ICs fabricated in a primary bipolar technology, and (ii) gate-coupled devices fabricated in an LDD, nonsilicided, 0.6 μm CMOS process.

3.1 STATISTICAL DISTRIBUTION OF EOS/ESD-RELATED FAILURES

The impact of EOS/ESD phenomena on different IC products was quantified using extensive failure analysis (FA) data compiled at Texas Instruments, Incorporated. The analysis of the database was done at different levels, namely, (i) product group, (ii) product subgroup, and (iii) device type. To illustrate the analysis done, two particular views of the database at the product level are presented. The impact of EOS/ESD phenomena on the failure rates for each product group was measured as the ratio between the number of EOS/ESD-related failures and the total number of failures. Figure 3.1 shows the percentage of the failures that can be attributed to EOS/ESD for various product groups. This figure reveals the relative magnitude of the EOS/ESD problem compared to assembly, wafer-fab, design, and other sources of failures. For example, Fig. 3.1 indicates that nearly 27% of the failures in the product family *A* were attributed to EOS/ESD. The database was also used to identify the percentage of EOS/ESD-related cases that received failure analysis (FA) per customer request. Figure 3.1 shows a typical result from this type of analysis at the product group level. For example, nearly 65% of the EOS/ESD failures in product family *A* received FA at customers' requests.

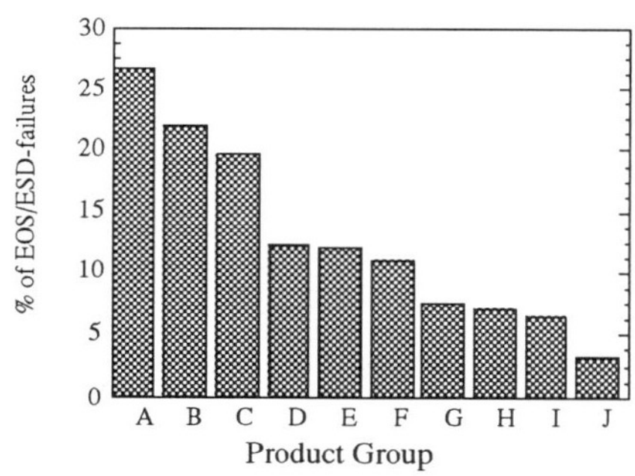

Figure 3.1 Relative impact of EOS/ESD phenomena on product group failure mechanisms.

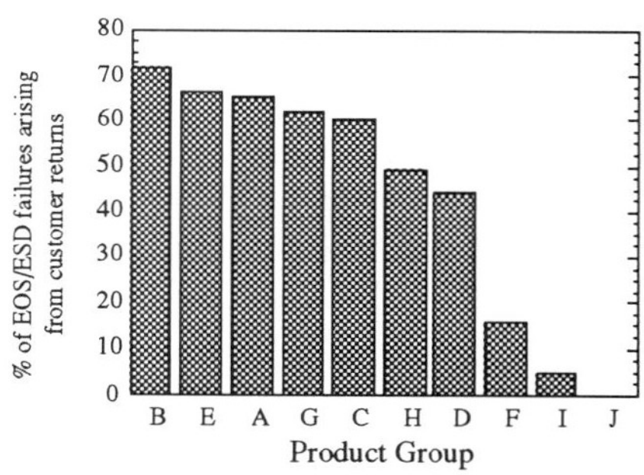

Figure 3.2 Percentage of EOS/ESD-related failure analysis (FA) jobs in the database that are attributed to customer requests.

Similar analyses were done at the device-type level. A ranking of products (by device type) was obtained by querying the database at the device level to identify those most affected by EOS/ESD. Four different device types (C1, C2, C3, and C4) from a representative product group (C) were chosen as test vehicles for this study. These devices were fabricated in a standard bipolar process. A schematic cross-section of a typical bipolar transistor in this process is shown in Fig. 3.3.

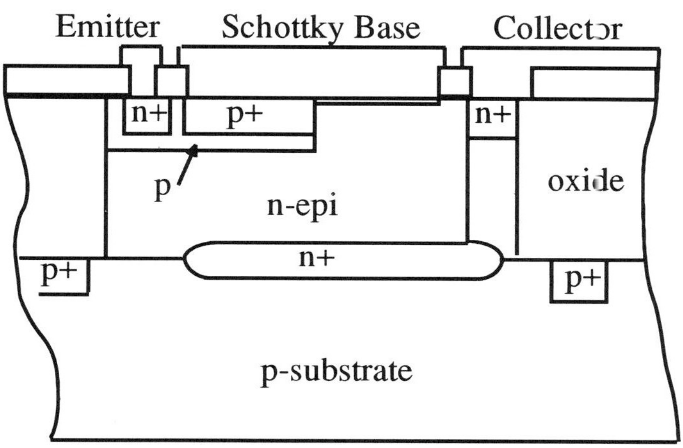

Figure 3.3 Schematic cross-section of a typical transistor for the bipolar technology used to fabricate the selected devices.

For the selected devices, database information revealed that junction burnout was the dominant EOS-related failure mechanism. The detailed FA reports were then analyzed to determine the particular failure modes. This information was then used to specify the stress method to reproduce the observed failure modes.

3.2 CHARACTERIZATION OF BIPOLAR DEVICES

3.2.1 Test Procedures

The electrical transients for the various EOS sources are difficult to characterize in the time domain or frequency domain. On the other hand, EOS square pulse voltage waveforms are easy to generate and simple to analyze [43, 76]. For these reasons, the square pulse was chosen as the EOS test waveform in our study. This type of EOS stressing was applied to a number of units for each selected device type. In order to compare the failure thresholds and failure characteristics for at least two different types of stress waveforms, HBM-ESD stress was considered. Several units of each selected device type were subjected to the HBM-ESD waveform stress.

Both EOS-square-pulse and HBM-ESD tests were performed only on unpowered devices, with all unstressed pins left floating. The stress polarity was always positive with respect to the *GND* pin. HBM-ESD experiments were conducted using a KeyTek tester. Square-pulse testing was done using either an HP214B pulse generator or a charged transmission line (TL) [63] as EOS sources. In particular, TL was used whenever high-current (more than 2 *A*) short-duration pulses were required.

Voltage and current waveforms at the stressed pins were recorded for both EOS and HBM-ESD to determine the power delivered to the pin under test (PUT) and the time to the onset of failure. Figure 3.4(a) shows the test setup for EOS stressing with a high power pulse generator. Figure 3.4(b) shows the typical response of an output pad to a 10 μs EOS pulse. The abrupt drop in the pad voltage, accompanied by a slight increase in the stress current, is a typical signature of device failure. In Fig. 3.4(b), the failure occurred around 5.8 μs with a stress power level of about 20 W. Device failure was defined as a curve trace (CT) softening or short to VCC or GND of the stressed pin characteristics. It is important to mention that each zap was followed by a CT evaluation of the pin even if no failure signature was detected in the stress waveforms.

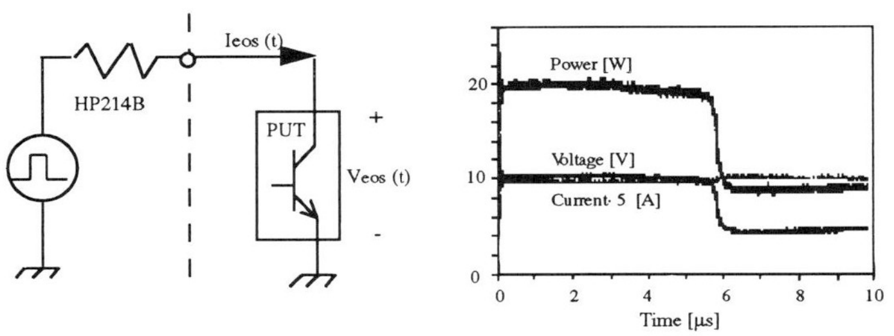

(a) (b)

Figure 3.4 A typical power profile characterization of an ESD/EOS protection transistor. (a) Experimental setup used for square-pulse EOS measurements. (For submicrosecond EOS stressing, transmission line pulsing was used.) (b) Oscilloscope traces of the pad voltage and current waveforms together with the computed power waveform for an output pin of the device C2 subjected to a 10 μs EOS event.

3.2.2 Experimental results

In this section, detailed experimental results (data-point level) are presented for two different representative device types. Also presented are the least-square fits of the experimentally derived profiles for all of the selected devices. As shown in Fig. 3.5, analytical thermal failure models [43] predict that there are four time domains in the pulse power-to-failure vs. time-to-failure profile. In particular, the $P_f \sim {t_f}^{-0.5}$ region is commonly referred as the Wunsch-Bell model [100].

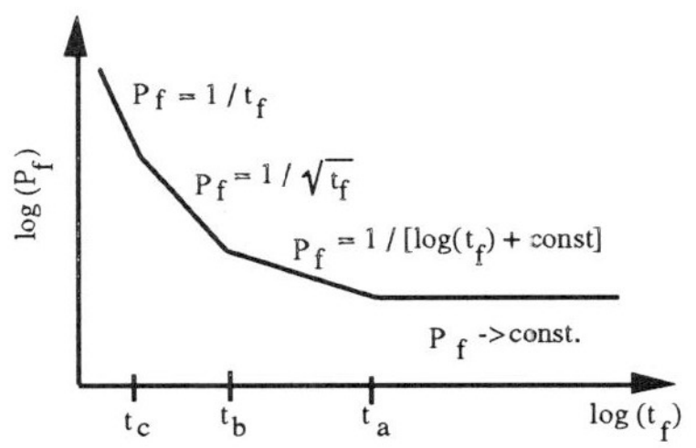

Figure 3.5 The four time domains of the power-to-failure vs. time-to-failure relationship.

Figure 3.2.2 illustrates a logarithmic plot of the power profile obtained for the output pulldown transistors of device type C2. The power profile clearly displays a nearly Wunsch-Bell domain together with the constant power domain.

Figure 3.2.2 shows the power profiles for two different layout design revisions of the lower output pulldown transistor in the bidirectional pins of device C1 (see Figs. 3.2.3 and 3.14 below). Figure 3.2.2 indicates that design revision 2 is more EOS resistant than revision 1 for stress levels with corresponding times-to-failure less than 20 μs. From the profiles extrapolated into the characteristic times of the HBM-ESD ($\log(0.15\ \mu s) \approx -0.82$), an improved ESD performance is expected for the layout design revision 2. This was corroborated by HBM-ESD experimental data presented below.

It is important to note that for all the stressed devices, it was difficult to obtain experimental data far beyond the beginning of the constant power regime. This can be explained in terms of the high-differential sensitivity of the time-to-failure with respect to changes in the stress power level within this particular regime. Slight variations in the power level favor a narrow distribution of failure

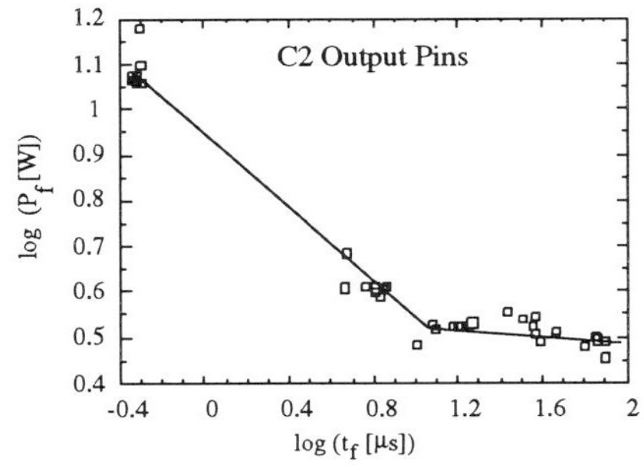

Figure 3.6 Power profile for the output pulldown transistor of device C2.

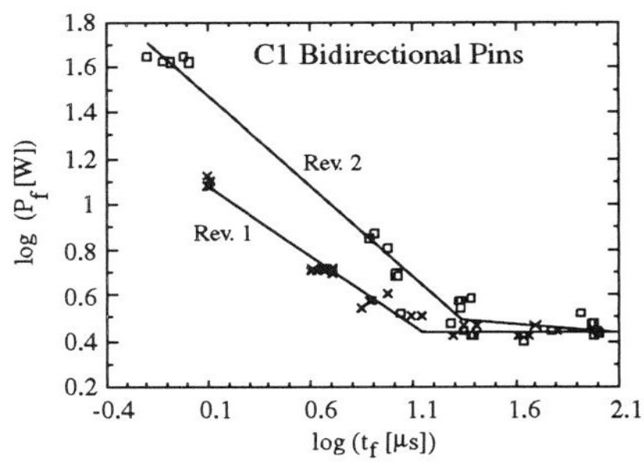

Figure 3.7 Power profiles of the output pulldown transistor in the bidirectional pins of device C1. Two different layout design revisions are considered here.

times that is only broadened by the random device-to-device variations and the parasitics of the test fixtures.

Experimental data of each power profile were approximated using a pair of functions generically written as

$$P_f = a t_f{}^b. \tag{3.1}$$

The least-square fit of the experimental data was used to determine the two sets of a, b coefficients required for each one of the four device types considered in this study.

Table 3.1 shows the results for both the input and output protection transistors of each device type. In this table, t_{fo} is the failure time at which the constant power domain of the power profile is approached. The plots of the fitted power profiles are shown in Figs. 3.2.2 and 3.2.2. The profile for the bidirectional pins of device C1 (Rev. 1) is included in both figures for reference. These fitted power profiles are accurate in the range of the experimental data, namely, from 0.4 μs to 100 μs. In Figs. 3.2.2 and 3.2.2, the power profiles have been extrapolated to the ESD regime. Table 3.2 shows ESD-HBM (positive with respect to ground only) experimental data for the devices under study. This table also includes the estimated ESD failure thresholds determined by extrapolating the EOS power profiles back into the ESD regime [42].

Some comments are in order regarding these profiles. First, it is clear from Table 3.1 that the outputs of devices C1.2 and the inputs of devices C4 show

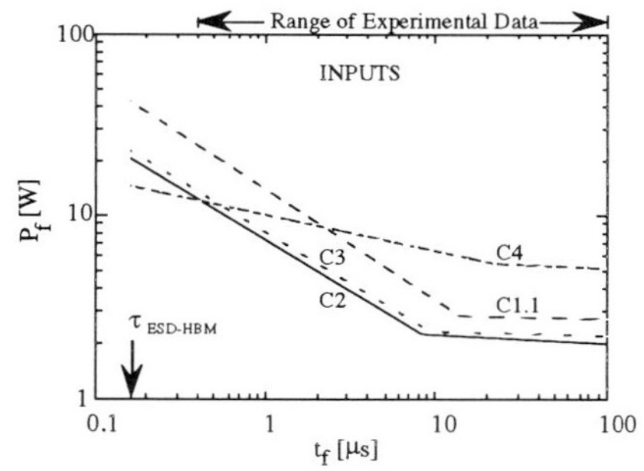

Figure 3.8 Power profiles for the input protection transistors. The arrow points to the HBM-ESD time constant (0.15 μs).

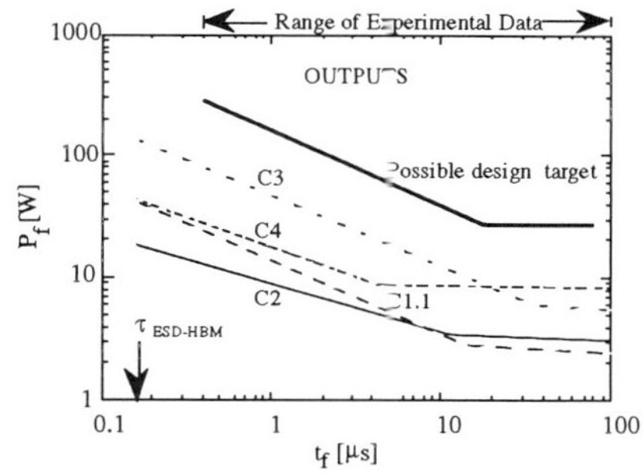

Figure 3.9 Power profiles of the output pulldown transistors.

the largest deviations from the typical Wunsch-Bell dependence where the parameter b is expected to be around -0.5. As shown by failure analysis, the stressed transistors in these pins exhibit more than one dominant failure mechanism thus explaining the observed power profile behaviour. Second, Figs. 3.2.2 and 3.2.2 provide an assessment of the relative resistance to EOS/ESD events by the different I/O circuits under the specified stress conditions. In particular, in view of the output pin EOS/ESD robustness, the pulldown transistors of device C3 require higher power levels to reach the onset of failure within time frames less than 20 μs. Third, comparison of output and input power profiles per device type indicates that the input protection circuits are generally more susceptible to the EOS/ESD damage than their output counterparts. This is primarily due to the fact that the dimensions of the input protection transistor are generally smaller than the corresponding output pulldown transistor dimensions in these particular devices. Finally, these power profiles suggest that high ESD failure thresholds *do not* warrant EOS robustness. This is particularly true for devices whose power profiles exhibit a large roll-off slope before reaching the constant power regime as is the case for the inputs of device C1.1. It can also be observed from data in Table 3.2 and Fig. 3.2.2 that although the ESD performances of the lower output pulldown transistors in devices C1.1 and C4 are similar, their EOS performances are different: the output transistors in C4 outperform those in C1.1 for the EOS regime.

The thick solid line in Fig. 3.2.2 exemplifies a *possible design target* for overall EOS/ESD device resistance. When the I/O protection circuits considered here undergo a design change aimed to improve EOS/ESD performance, the effectiveness of the design change can be evaluated by checking how much the power profile of the revised version can be moved towards the design target.

3.2.3 Failure Analysis

Two main objectives were pursued with the FA work. They were (i) to determine the ability of the test procedure to reproduce the observed dominant failure mechanisms found in the selected device types, and (ii) to determine any tendencies in the EOS and ESD failure mechanisms in terms of layout style and stress waveform parameters. For each device type considered, several units were submitted to FA to assure statistically valid results.

As expected, EOS/ESD damage site locations were layout dependent. For layout styles such as the one shown in Fig. 3.2.3, both EOS and ESD damage locations were generally located on the emitter contact edge facing the collector contact. ESD failures showed up as small damaged areas located exclusively at the corners of the emitter contact, while EOS failures appeared as relatively large spots with the specific location dependent on the stress level used. The failure site location moved away from the corner along the said emitter edge as the stress level decreased (time-to-failure increased). Under low-level long-duration stress, there is time for more uniform distribution of the stress across the transistor area. It is important to point out that failures involving the collector contact and the grounded guard ring were also seen in transistors with a similar layout style but of different dimensions.

Figure 3.2.3 shows an SEM photograph of a typical input protection circuit for device C4 that shows this type of failure. It was found that whenever more than one failure mode existed in a given transistor, the corresponding power profile departed greatly from the Wunsch-Bell relation. This was the case for the output pulldown transistor of device type C1 and the input protection transistor of device type C4.

For the transistor layout shown by the SEM photograph in Fig. 3.2.3, the failure sites were always located at the indicated emitter finger tip regardless of the stress type and duration. Figures 3.2.3 and 3.14 show SEM photographs of the lower output pulldown transistors in the bidirectional pins of device C1 in design revisions 1 and 2, respectively. In C1.1 output transistors, the failure site was always found at either one of the corners of the U-shaped emitter stripe as shown in Fig. 3.2.3. For devices C1.2, no consistent failure sites were found along the emitter stripe. Note that the avoidance of the U-

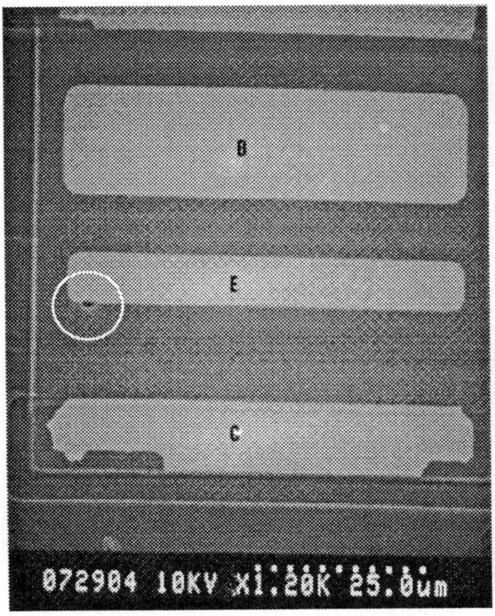

Figure 3.10 SEM photograph of the output pulldown transistor of the C2 device showing the typical failure site location.

Figure 3.11 SEM photograph of a typical input protection transistor for device C4 showing one of the two dominant failure site locations.

shaped emitter contact and the increase in the emitter to collector-contact spacing implemented in C1.2 resulted in an improved ESD/EOS performance as measured by the power profiles shown in Fig. 3.2.2. From the failure location patterns, the relative EOS performance and the size of the output transistors shown in Figs. 3.2.3 and 3.2.3 in comparison with the output transistors of devices C1.2 (Fig. 3.14, more conventional layout style), one can conclude that the EOS performances of the output pins in devices C1.1 and C4 are limited due to their layout style. It also suggests that there should be an EOS design guideline on the emitter-contact to collector-contact spacing

We conclude this section by noting that the suitability of the proposed test procedure was confirmed by FA; the test-induced failures matched the field failures. Furthermore, the EOS/ESD performance of those devices, measured in terms of the power profiles shown in Figs. 3.2.2 and 3.2.2 was qualitatively explainable based on the detailed FA results and the layout information.

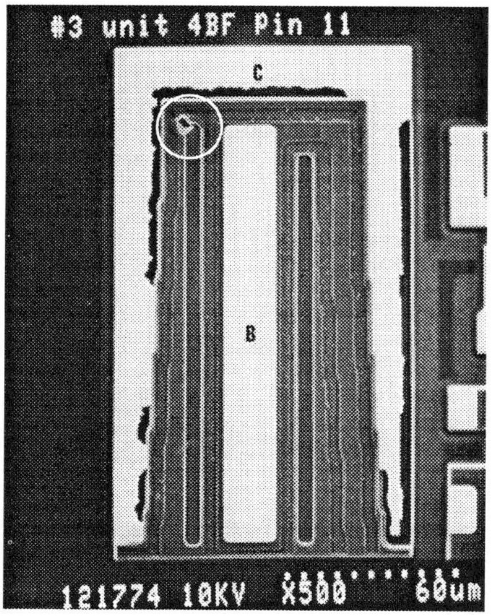

Figure 3.12 SEM photograph of the output pulldown transistor of a C4 device showing EOS failure site location. Devices with this layout style and dimensions were found to fail at the emitter finger tip closer to the collector contact.

Figure 3.13 SEM photograph of the output pulldown transistor (revision 1) in a C1 device showing EOS failure site location. For this layout style, the failure sites were always found at either one or both corners in the U-shaped emitter stripe.

3.3 EOS CHARACTERIZATION OF NMOS DEVICES

3.3.1 Device Description and Test Procedure

The structures studied in this section used 0.6 μm poly gate transistors with 110 Å gate oxide. The process, similar to the one described in [12], involved a boron punch-through implant followed by a phosphorus LDD implant and arsenic source-drain implants. A typical nMOS device cross-section and simulated doping profiles are shown in Fig. 3.15 for this technology. As seen, the effective junction depth is nearly 0.37 μm for this technology.

Grounded-gate multifinger nMOS devices are known to have a broad ESD failure-threshold distribution [77]. This has been shown to be the result of

Figure 3.14 SEM photograph of the output pulldown transistor (revision 2) in a C1 device showing EOS failure location.

stress current lock-on in a reduced number of device fingers, possibly due to finger-to-finger variations. Figure 3.16(a) is an emission microscopy picture showing the lock-on phenomenon in a multifinger gate-grounded device being stressed with a 1 kHz, 500 mA current pulse train [35].

The ESD/EOS experiments presented here were done using the novel gate-coupled multifinger nMOS (GCNMOS) structure introduced in Section 2.5 (see also [34]) and schematically shown in Fig. 3.17(a). The purpose of the field oxide device (FOD) is to allow transient MOS conduction of the thin-oxide protection device during the first moments of high-level ESD stress events. Appropriate levels of gate coupling generate enough substrate current that together with sufficient of drain ballasting resistance (proportional to the drain contact-to-gate spacing) facilitate the uniform turn on of all fingers in a multifinger device as shown in Fig. 3.16(b). The effect of the drain ballasting resistance can be seen in the *I-V* characteristics shown in Fig. 3.17(c) where the arrows show the successive turn-on of device fingers. Increasing the drain ballasting resistance reduces the required current swing between finger turn-on levels. Uniform power distribution by preventing current lock-on phenomenon has been shown to result in improved ESD performance [34, 35].

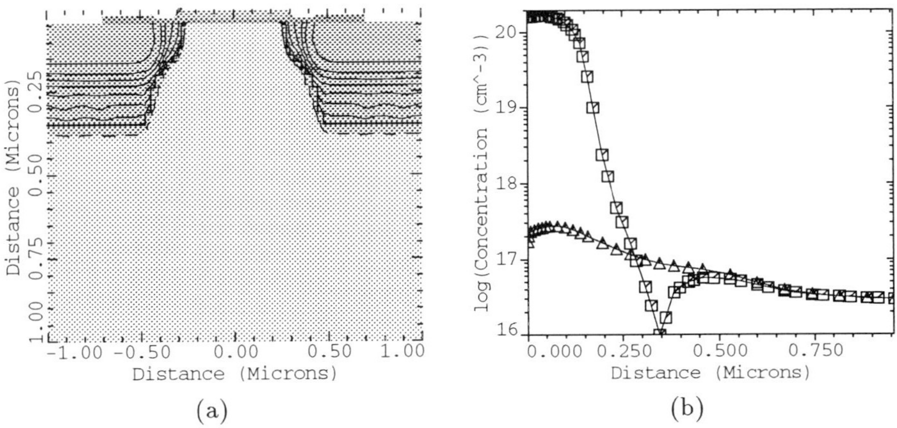

Figure 3.15 (a) LDD nMOS transistor cross-section and (b) the midchannel and drain/source doping profiles.

Gate-coupled devices with varying width (W) and contact-to-gate spacing (CGS) were fabricated in the LDD nonsilicided 0.6 μm process described above. Devices with different widths and drain/source contact-to-gate spacings were characterized for EOS using unipolar positive pulses applied to the pad. The failure criterion was defined as a change in the device leakage characteristics. Current pulses generated by a high-power pulse generator or a charged transmission line [63] were used for EOS stressing; a schematic of the experimental setup is shown in Fig. 3.18. With a fixed pulse width, the level of stress current was increased between current pulse-zaps until failure signatures were observed either in the waveforms or in the leakage level. The stress current level, the average power required to cause failure, and the time-to-failure were recorded and used to generate the current-to-failure and power-to-failure vs. time-to-failure profiles.

Figure 3.3.1(a) shows experimental power profiles for various devices having equal contact-to-gate spacings (CGS) but different widths. Because of the uniform conduction, increasing the device width improves the overall EOS/ESD performance. However, this figure also indicates that the degree of performance improvement is dependent on the stress duration and that only for long times-to-failure does it tend to the ideal scaling given by $W_2 \times P_f(t_f, W_2) = W_1 \times P_f(t_f,\ W_1)$, where P_f and t_f are the power required to cause failure and the time at which failure occurs, respectively. This can be explained if one considers

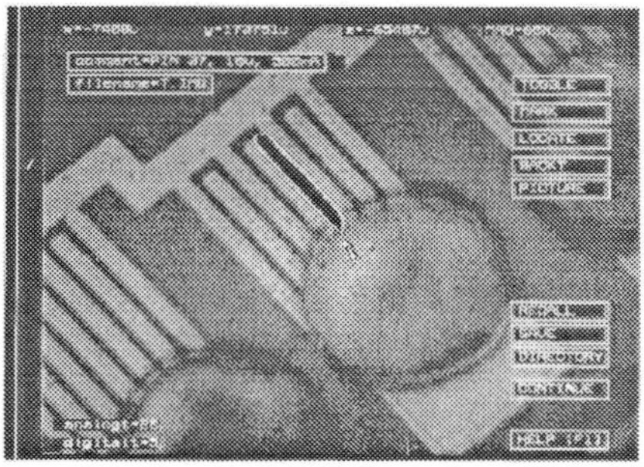

(a)

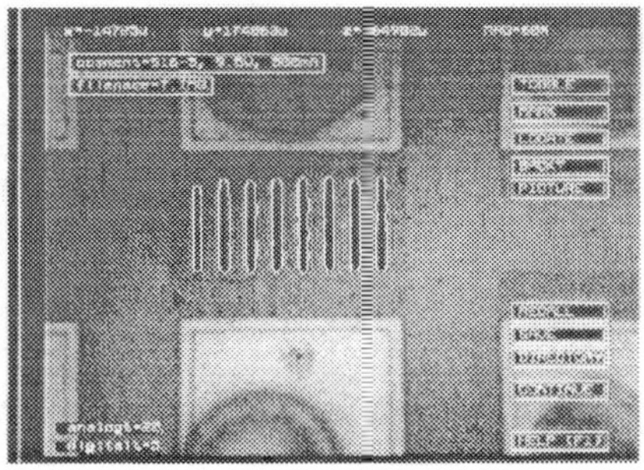

(b)

Figure 3.16 (a) Current lock-on phenomenon in a multifinger gate-grounded device while subjected to a 1 kHz, 500 mA current pulse train. (b) Uniform conduction in a multifinger gate-coupled device for the same stress conditions.

that long times-to-failure involve low stress power levels, thus allowing the devices to reach a more uniform stress energy distribution. In contrast, geometry variations across the device might exacerbate nonuniform current distribution under high power ESD/EOS stress events.

Figure 3.3.1(b) shows the experimental power profiles for a 500 μm wide device for various values of the drain contact-to-gate spacing. As can be seen, the increase in DCGS improves the power handling capability suggesting a more robust design for EOS.

The experimental stress current vs. time-to-failure (current profiles) are shown in Figs. 3.20(a) and 3.20(b). Figure 3.20(b) indicates that the current profiles have a low sensitivity with respect to CGS. This suggests that the improved power handling capability seen in Fig. 3.20(b) can be attributed mainly to the additional voltage drop across the drain and source diffusions as a result of the **uniform** contact-to-gate spacing increase.

3.3.2 Failure Analysis

Stressed devices were submitted for failure analysis. The SEM photographs shown in Fig. 3.21 are representative of the failures found in all stressed devices. Figure 3.21(a) shows an SEM photograph of a 750 μm wide device stressed with a 5 A, 1.0 μs current pulse. The photograph reveals that the device failed because of the formation of a poly filament in the drain side in one of the

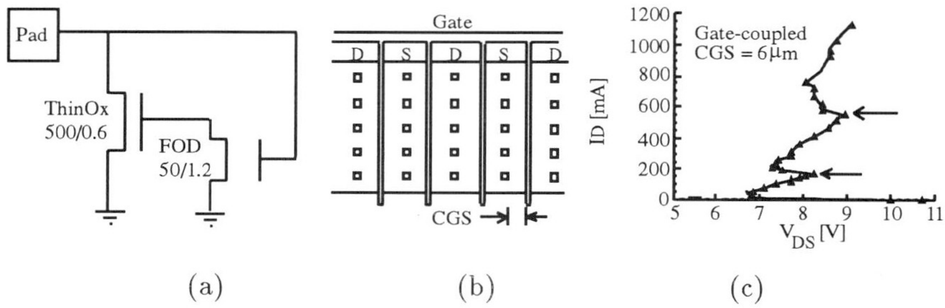

Figure 3.17 (a) Circuit schematic of a gate-coupled ESD protection structure, (b) schematic multifinger layout and, (c) typical *I-V* characteristics.

fingers of the nMOS device. Figure 3.21(b) is an SEM photograph showing the damage in a 500 μm wide device stressed with a 1.5 A, 10 μs current pulse; the oscilloscope traces indicated that this device failed at 8.7 μs. Figure 3.21(c) shows the damage in a 750 μm wide device stressed with a 1.5 A, 100 μs current pulse; the oscilloscope traces indicated that this device failed at 8.7 μs. As the stress duration increased, the damage site size always increased. This is due to the increasing levels of energy that can be delivered to the failure site if the device fails early during the stress event. Another factor contributing to the failure site size is that device heating is more uniform under low-level stress conditions.

3.4 SUMMARY

An industrial failure analysis database was used to evaluate the relative impact of EOS phenomena on device failure across the different product families. The use of database information allowed us to identify the dominant EOS-related failure mechanisms affecting different products. A set of different devices was selected as test vehicles for this analysis. A test procedure was specified and used to consistently reproduce the dominant failure mechanisms observed in the field for the selected set of devices. Two kinds of unipolar stresses were used, square-pulse testing and HBM-ESD. Stressing a number of units for each selected device type allowed us to determine the power profiles for individual devices. Detailed FA was done on stressed units to determine the failure mechanisms and to provide information useful to the qualitative explanation of the differences observed among the measured power profiles.

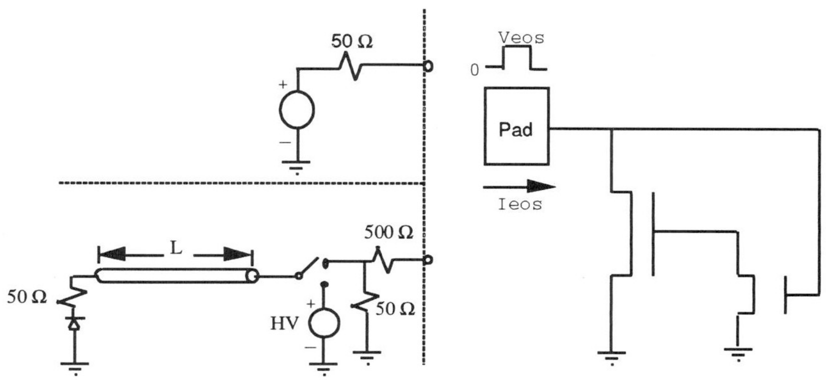

Figure 3.18 The EOS testing environment. Circuit schematic showing the gate-coupled protection structure and the sources of EOS: a high power pulse generator or a charged-transmission line (HV stands for high voltage supply).

A good qualitative correlation was found between the layout style and the device ability to sustain EOS/ESD stress as measured by the power and current profiles (device failure characteristics). It was shown that good IC ESD-HBM performance does not warrant adequate EOS robustness. Design changes of I/O protection structures can be evaluated in terms of the device failure characteristics. Relative device EOS performance, as indicated by the device failure characteristics, can also be used to set design revision priorities.

FA results showed that the specified test procedure can consistently reproduce the observed dominant failure mechanisms. The differences seen in the measured power profiles were explained through FA in terms of the failure location, layout style, and stress level. FA also revealed tendencies in both the damage site size and damage site location in terms of the time-to-failure for certain layout styles.

The results presented suggest that current profiles can be used to set a design goal for EOS and ESD of I/O protection circuits. Furthermore, device failure characteristics derived from experimental data and/or accurate device electrothermal models together with FA information may be used to develop design guidelines that incorporate EOS/ESD performance specifications.

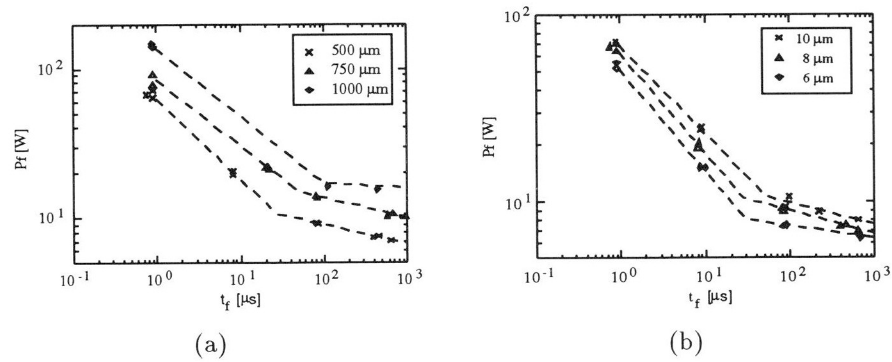

Figure 3.19 Experimental power profiles: (a) devices with $CGS = 8\,\mu m$ and different widths, and (b) 500 μm wide devices with different contact-to-gate spacings.

Table 3.1 Power profile coefficients.

Device Type	Pin Type	t_{fo} [μs]	$t \leq t_{fo}$		$t \geq t_{fo}$	
			a [$W/\mu s^b$]	b	a [$W/\mu s^b$]	b
C1.1	Bidir	13.8	13.8	0.61	2.8	0.003
C1.2	Bidir	20.4	35.5	0.8	4.0	0.08
C2	Input	8.2	7.3	0.56	2.5	0.05
	Output	11.5	8.9	0.4	3.7	0.04
C3	Input	9.2	8.1	0.56	2.5	0.03
	Output	34.8	47.6	0.58	9.3	0.12
C4	Input	22.1	10.1	0.2	6.1	0.04
	Output	4.25	17.8	0.5	8.7	0.005

Table 3.2 HBM-ESD failure thresholds.

Device Type	Input Pins		Output/Bidir Pins	
	Measured [kV]	Estimated[a] [kV]	Measured [kV]	Estimated[a] [kV]
C1.1	10.5		6.5	6.9
C1.2	10.5		> 10	
C2	3.2	3.0	3.5	2.7
C3	2.7	3.4	9.5	12.1
C4	2.0	1.0	7.0	8.6

[a] Assuming current follows a similar relationship as the power does.

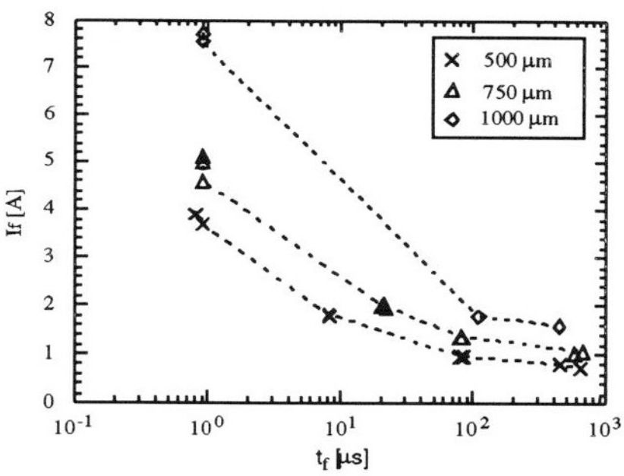

(a)

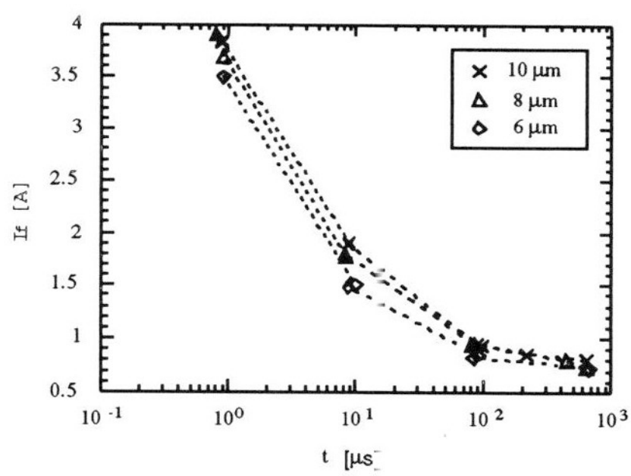

(b)

Figure 3.20 Experimental current profiles: (a) devices with $CGS = 8\ \mu m$ and different widths, and (b) 500 μm wide devices with different contact-to-gate spacings.

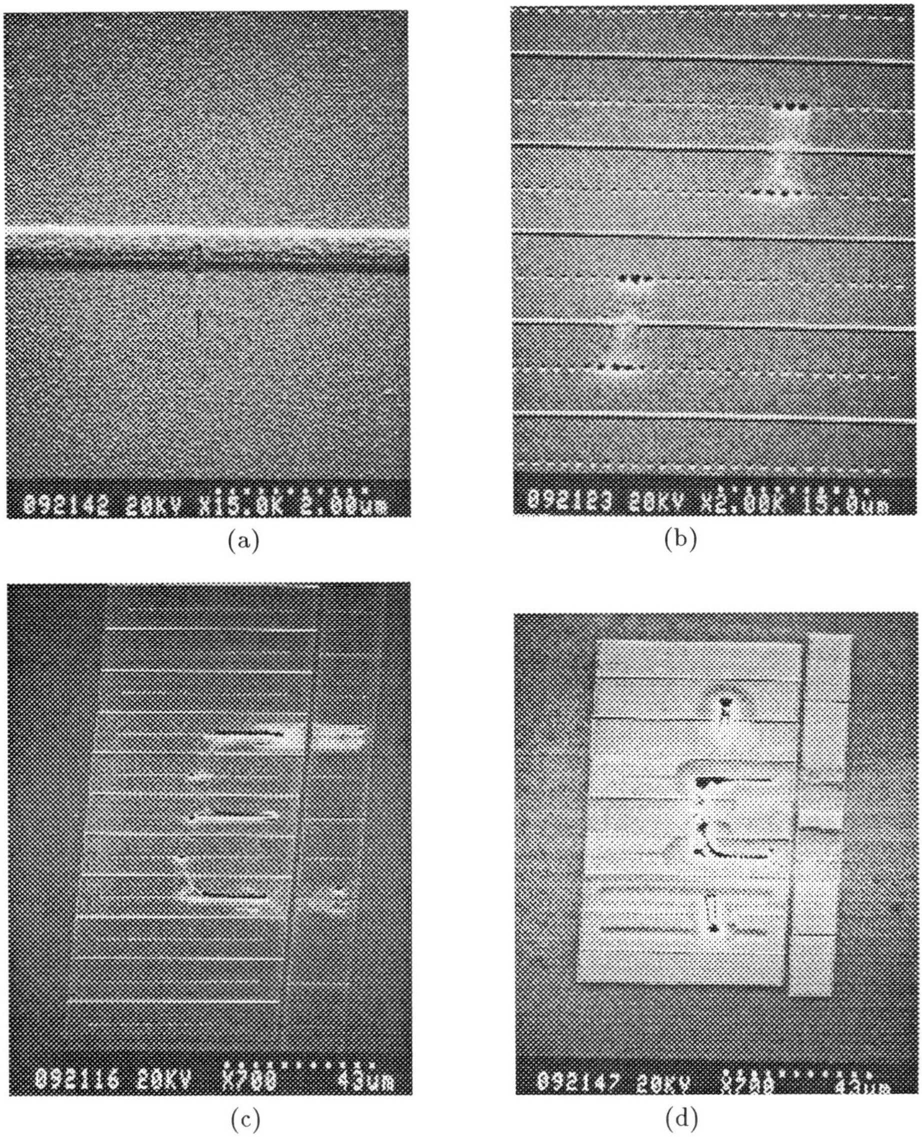

Figure 3.21 SEM photographs of gate-coupled devices after EOS current-pulse zapping: (a) a poly filament in a 750 μm wide, $CGS = 8\ \mu m$ device stressed with a 1.0 μs, 5 A current pulse; (b) drain-source damage seen in a 500 μm wide, $CGS = 6\ \mu m$ device stressed with a 10 μs, 1.5 A current pulse; (c) damage in a 750 μm wide, $CGS = 8\,\mu m$ device subjected to a 100 μs, 1.5 A current pulse; (d) gross failure in a 500 μm wide, $CGS = 8\ \mu m$ device after a 1 ms, 0.8 A current pulse.

4

EOS THERMAL FAILURE SIMULATION FOR INTEGRATED CIRCUITS

In this chapter, a nonlinear thermal simulator, iTSIM, is introduced. New EOS/ESD thermal models have been developed and implemented in iTSIM to overcome the limitations found in previous analytical thermal failure models. Section 4.2 gives an overview of iTSIM, its main features and simulation results. An efficient mixed *2D-1D* simulation approach is used in iTSIM to handle thermal failure conditions associated with EOS events for which the scale of the thermal problem is large (on the order of the wafer thickness). Appropriate boundary conditions are derived and implemented to support the mixed *2D-1D* simulation capability. Simulation results for ICs in ceramic and plastic packages are presented in Section 4.3. Numerically derived power profiles are compared with experimental data [24, 23]. Finally, the conclusions are presented in Section 4.4.

4.1 NOMENCLATURE

- $Arg()$ – principal value of the complex function arg().
- C_A^* – heat capacity per unit area of the package material in contact with the passivation oxide.
- C_M^* – heat capacity per unit area of the die-bond and the package material in the wafer backside.
- $H_i(t)$ – impulse response of the thermal impedance Z_i.
- $(HoQ)_{i,n}$ – value of the convolution integral for grid node i on boundary 1-3 at time t_n.

- Q_i – heat flowing into boundary grid node i.
- R_A^* – thermal resistance per unit area of the package material in contact with the passivation oxide.
- R_m^* – thermal resistance per unit area of the die-bond and the package material in the wafer backside.
- T_i – temperature at grid node i.
- Z_i – thermal impedance at grid node i on boundary 1-3.
- c_{p_ox}, c_{p_s} – silicon dioxide and silicon specific heat.
- t_{ox} – passivation oxide layer thickness.
- $x_{i,n}^{(k)}$ – value of the variable x at grid node i in the kth iteration of Newton's method at time t_n.
- κ_{ox}, κ_s – silicon dioxide and silicon lattice thermal conductivities.
- ρ_{ox}, ρ_s – silicon dioxide and silicon densities.
- ζ_i – steady-state thermal resistance between boundary node i and the ambient.

4.2 ITSIM: A NONLINEAR THERMAL FAILURE SIMULATOR FOR ICS

iTSIM, written in C language, performs nonlinear steady-state and transient analyses. The nonlinear heat diffusion equation is discretized using a finite difference scheme. For transient analysis, iTSIM uses either backward Euler or trapezoidal integration methods with fixed and/or variable time steps. The nonlinear system of equations is solved using Newton's method with a sparse matrix solver [57] for the solution of the linearized system. Adiabatic (Neumann) and ohmic-like boundary conditions (BCs) are allowed in iTSIM. Input specification for iTSIM is similar to that for the device simulator PISCES-IIB. The program handles three types of simulation domains, namely, *1D*, pure *2D* with reflective BCs, and mixed *2D-1D* with nonadiabatic BCs. For EOS failure modeling, failure is defined as the time at which a given structure reaches a predefined critical temperature such as silicon melting or aluminum melting. During simulation of EOS failures, *iTSIM* searches for the user-specified failure criterion along the principal axis of the heat source (defect site) at each time step. *iTSIM* stops the simulation when the failure condition is not reached in a reasonable amount of time by monitoring the rate of temperature change at the grid nodes.

Two typical simulation domains for iTSIM are shown in Figs. 4.1 and 4.2. Since the problem is symmetrical around the principal axis of the heat source (defect) perpendicular to the Si-SiO2 interface, only half of the structure is actually simulated by using adiabatic boundary conditions along that heat source's principal axis. In Fig. 4.1, adiabatic boundary conditions are also used over the left boundary of the simulation domain. This selection of the boundary conditions is suitable for a periodic array of heat sources spaced by a distance equal to two times the width of the simulation domain. In EOS/ESD studies, one is generally interested in thermal failures due to a single heat source; in this case iTSIM uses a simulation domain as shown in Fig. 4.2. Since EOS events could have characteristic times as high as several milliseconds, the heat wavefront has enough time to reach points as far as 200 μm away from the heat source. Since the *2D* simulation domain of that size is difficult to handle without compromising the accuracy of the results and the stability of the algorithms, iTSIM uses a mixed *2D-1D* simulation approach. The *2D* spatial discretization is explicitly considered only for the region containing the heat source (cross-hatched) that is bounded on the left by the user-specified surface boundary $2-1-3$ in the dark solid line in Fig. 4.2. The rest of the die plus the package material to the left of this rounded boundary is divided into two regions, each of which is modeled by the appropriate *1D* BCs introduced in the next section.

4.2.1 Boundary Condition Models for EOS Thermal Modeling

As mentioned earlier, the adiabatic BCs on the left side of the 2D structure shown in Fig. 4.1 are suitable for analysis of periodic arrays of defects spaced by a distance equal to two times the simulation domain width. This section shows how to provide a computationally efficient environment suitable for *2D* simulation of IC thermal structures. The key is in the development of a mixed-$2D/1D$ simulation capability. The *2D* domain is terminated by two user-specified boundaries beyond which the structure is represented by two *1D* distributed models as depicted in Fig. 4.2. If the solution to the steady-state problem were known a priori, this boundary would correspond to an isothermal surface. Since this is not the case, an assumption has to be made regarding the shape of this surface, in particular, for the part of the problem associated with the region below the horizontal axis of the defect. If the simulation domain size

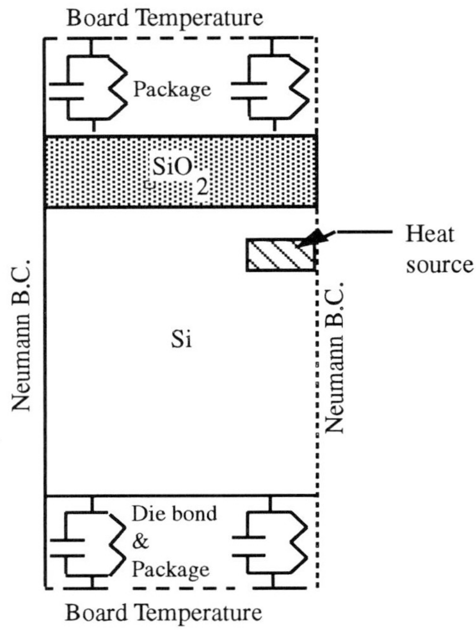

Figure 4.1 Schematic of a pure *2D* simulation domain for iTSIM.

is much larger than the defect dimension, one can assume that this surface can be approximated by an arc shown in Fig. 4.2.

To address the thermal BC problem, it is assumed that the structure can be divided into two parts which correspond to the regions divided by the principal axis of the defect parallel to the Si-SiO_2 interface. The upper region is bounded on the left by the 1-2 surface. The lower region is bounded by the 1-3 surface and the vertical principal axis of the defect as shown in Fig. 4.2.

Let us first consider the upper portion of the structure, i.e., the grid nodes on the boundary *1-2* as depicted in Fig. 4.2. To compute the thermal impedance associated with these nodes, the die plus the package portion of the structure beyond the 1-2 boundary is approximated by a distributed RC network as shown in Fig. 4.3(a). If the heat flowing into node i of this boundary is Q_i, and the grid spacing associated with node i is l_i, one can show that the thermal

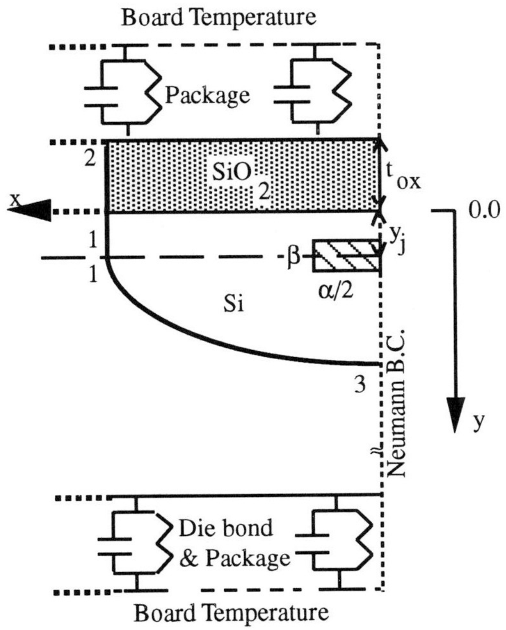

Figure 4.2 Schematic of a mixed *2D-1D* simulation domain for iTSIM. Pure *2D* simulation is constrained to the region bounded by the vertical principal axis of the defect and the user-specified boundary *2-1-3*.

impedance is given by

$$Z_i = \frac{w}{l_i}\sqrt{r_p r_s}\sqrt{\frac{1}{1+sr_pC_p}} \tag{4.1}$$

with

$$r_p = \frac{t_{ox}}{\kappa_{ox}} + R_A^* \tag{4.2}$$

$$r_s = \frac{1}{\kappa_s Y_J} \tag{4.3}$$

$$C_p = C_A^* + \rho_{ox} c_{p_{ox}} t_{ox} + \rho_s c_{p_s} Y_j \tag{4.4}$$

for grid nodes in Si, and

$$r_p = R_A^* \tag{4.5}$$

$$r_s = \frac{1}{\kappa_{ox} t_{ox}} \tag{4.6}$$

$$C_p = C_A{}^* + \rho_{ox} c_{p_{ox}} t_{ox} \tag{4.7}$$

for grid nodes in SiO_2.

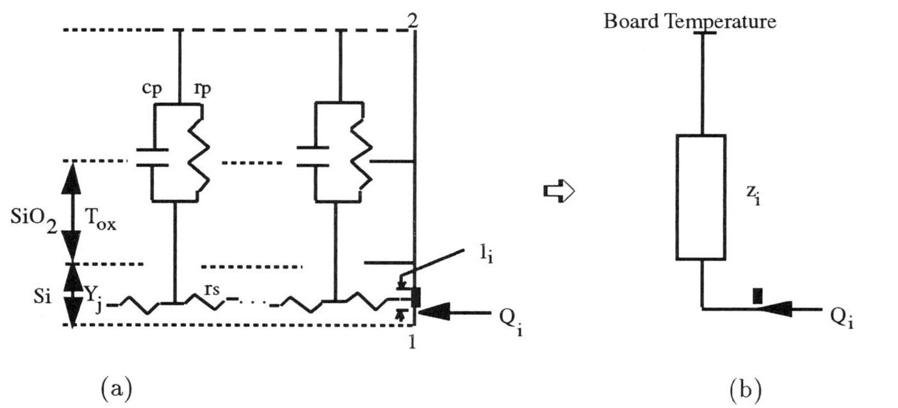

Figure 4.3 Schematic of the modeling approach used for the boundary conditions at grid nodes on the *1-2* boundary. (a) Distributed model schematic and (b) lumped thermal model for the boundary condition at node i.

In these expressions, the silicon dioxide thermal parameters and the silicon's heat capacity are held constant and evaluated at the user-specified board temperature. This is a good approximation given the weak dependencies of these parameters on the temperatures of interest. The silicon thermal conductivity is not held fixed. Instead, it is assumed that κ_s associated with a particular grid node in the 1-2 boundary is controlled by the the node's temperature and the user-specified board temperature. Actually, at each iteration, κ_s is linearized around an average temperature given by $< T_i >= DT_i + (1 - D)T_{board}$, where D is a user-specified parameter in the interval [0,1].

Initially, each Z_i was approximated in iTSIM by a second-order network connected to the corresponding boundary grid node. However, it was found that, under this implementation, simulated power profiles could predict contradictory trends for certain time-to-failure time intervals. To solve this problem, a full implementation of Z_i based on convolution techniques was used for transient analysis. At each grid node, the impulse response of Z_i is convolved in time with the node's heat flow to determine the corresponding stamp entries.

For consistent implementation of these BCs for both steady-state and transient analysis, each Z_i is implemented as a current-controlled voltage source. The stamp for this kind of voltage sources is written as [1]

$$\begin{vmatrix} 0 & 1 \\ mv^{(k)} & mi^{(k)} \end{vmatrix} \begin{vmatrix} {v_i}^{(k)} \\ {I_j}^{(k)} \end{vmatrix} : \begin{vmatrix} 0 \\ rhs \end{vmatrix} \tag{4.8}$$

where (k) denotes the kth iteration of Newton's method.

For steady-state analysis, the nodal temperature is related to the thermal resistance of the node under consideration and the heat flowing into it by

$$T_i = \zeta_i(T_i)Q_i + T_{board} \tag{4.9}$$

where Q_i stands for the heat flow into grid node i at the boundary and $\zeta_i = \frac{w}{l_i}\sqrt{r_{p,i}r_{s,i}}$ is an implicit function of the nodal temperature T_i through $r_{p,i}$ and $r_{s,i}$. Thus one obtains

$$mv^{(k+1)} = \frac{\partial \zeta^{(k)}}{\partial {T_i}^{(k)}}{Q_i}^{(k)} \tag{4.10}$$

$$mi^{(k+1)} = -{\zeta_i}^{(k)} \tag{4.11}$$

$$rhs^{(k+1)} = T_{board} - {Q_i}^{(k)}\frac{\partial {\zeta_i}^{(k)}}{\partial {T_i}^{(k)}}{T_i}^{(k)} . \tag{4.12}$$

On the other hand, for transient analysis, the impulse response $H_i(\tau)$ of Z_i can be written as

$$H_i(\tau) = \frac{\zeta_i}{\sqrt{\pi a_i \tau}} e^{-\frac{\tau}{a_i}} \tag{4.13}$$

with $a_i \equiv r_{p,i}c_{p,i}$. Therefore, the transient response $T_i(t)$ at node i can be computed according to

$$T_i(t) = \zeta_i(HoQ)_i + T_{board} \tag{4.14}$$

where $(HoQ)_i$ is the convolution integral for grid node i on boundary *1-3* at time t defined as

$$(HoQ)_i \equiv \int_0^t \frac{H_i(\tau)}{\zeta_i} Q_i(t-\tau)d\tau . \tag{4.15}$$

With the change of variable $x^2 = \frac{\tau}{a}$, one obtains

$$T_i(t) = \frac{2\zeta_i}{\sqrt{\pi}} \int_0^{\sqrt{t/a_i}} Q_i(t - ax^2)e^{-x^2}dx + T_{board} . \tag{4.16}$$

If $f(x)$ is an integrable function over $[x_0, x_n]$, then for a fine enough partition x_j of the integration interval, the following approximation is valid:

$$\int_{x_0}^{x_n} f(x)dx \approx \sum_{j=1}^{n} \frac{1}{2}(f(x_j) + f(x_{j-1}))(x_j - x_{j-1})\,. \tag{4.17}$$

This integration scheme is used to approximate the value of the convolution integral. For minimum computational overhead, the partition of the integration interval was chosen such that interpolation could be avoided. The time steps determined by the simulator are used for partitioning as shown schematically in Fig. 4.4. This partition guarantees that heat flux values required in the summation are directly available from the simulation results and need not be interpolated. To determine the stamp entries for the voltage source that models the transient response of node i at time t_n, $(HoQ)_i$ is rewritten as

$$(HoQ)_{i,n} = \eta_i Q_i(t_n) + \theta_{\mathbf{i,n}} \tag{4.18}$$

where

$$\begin{aligned}
\theta_{\mathbf{i,n}} &\triangleq \eta_i Q_i(t_n) + \eta_{\mathbf{i}} \mathbf{Q_i}(\mathbf{t_n} - \tau_{\mathbf{1}})\mathbf{e}^{-\tau_{\mathbf{1}}/\mathbf{a_i}} + \frac{\mathbf{1}}{\sqrt{\pi \mathbf{a_i}}} \sum_{\mathbf{j=2}}^{\mathbf{n}} \left[\sqrt{\tau_{\mathbf{j}}} - \sqrt{\tau_{\mathbf{j-1}}}\right] \\
&\quad \times \left[\mathbf{Q_i}(\mathbf{t_n} - \tau_{\mathbf{j}})\mathbf{e}^{-\tau_{\mathbf{j}}/\mathbf{a_i}} + \mathbf{Q_i}(\mathbf{t_n} - \tau_{\mathbf{j-1}})\mathbf{e}^{-\tau_{\mathbf{j-1}}/\mathbf{a_i}}\right] \\
\eta_i &\triangleq \sqrt{\frac{\tau_1}{\pi a_i}}.
\end{aligned}$$

In Eq. (4.18), $\theta_{i,n}$ is a known quantity when solving for the simulation time point t_n. Therefore, considering the heat flow $Q_i(t_n)$ as the independent variable, the nodal temperature for the kth iteration of Newton's method can be written as

$$T_i{}^{(k)}(t_n) = \zeta_{i,n}^{(k)}(HoQ)_{i,n}^{(k)} + T_{board} = \zeta_{i,n}^{(k)}\eta_i Q_i^{(k)}(t_n) + \zeta_{i,n}^{(k)}\theta_{i,n} + T_{board} \tag{4.19}$$

from which the stamp (matrix) entries can be immediately written as

$$\begin{aligned}
mv^{(k+1)} &= 1 - \frac{\partial \zeta_{i,n}}{\partial T_{i,n}{}^{(k)}}\left[\eta_i Q_i^{(k)}(t_n) + \theta_{i,n}\right] \\
mi^{(k+1)} &= -\zeta_{i,n}^{(k)}\eta_i \\
rhs^{(k+1)} &= T_{board} + \zeta_{i,n}^{(k)}\theta_{i,n} - \frac{\partial \zeta_{i,n}}{\partial T_{i,n}{}^{(k)}}\left[\eta_i Q_i^{(k)}(t_n) + \theta_{i,n}\right] T_{i,n}{}^{(k)}\,.
\end{aligned} \tag{4.20}$$

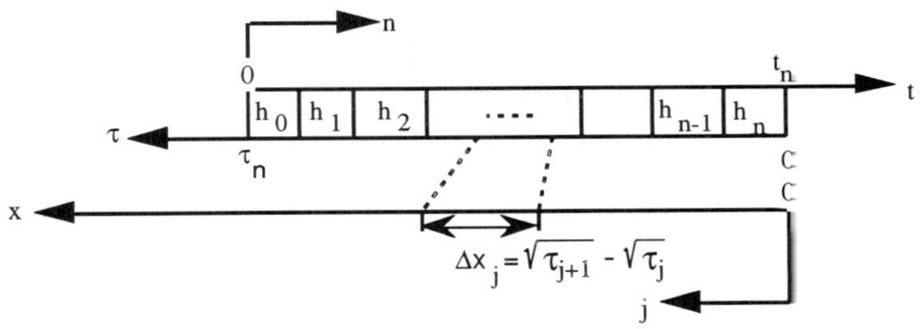

Figure 4.4 Partitioning of the integration interval according to simulation time steps.

Note that the approximation of the convolution integral by a summation naturally leads to the creation of the stamp entries that, even for the case in which r_s is a nonlinear function of the local grid node temperature, results in a straightforward implementation of Newton's method.

Let us now consider the grid nodes on boundary *1-3* shown in Fig. 4.5(a); we assume that the boundary *1-3* is isothermal for the structure under study and that this boundary can be approximated by an arc of circumference. Then, one can compute the equivalent lumped resistance and the heat capacity associated with the material bounded by the *1-3* surface and the backside contact of the wafer. To this end, we consider the equivalent problem shown in Fig. 4.5(b). This structure can be mapped into the one depicted in Fig. 4.5(c) by the sequential application of the following conformal mapping transformation [8]:

$$\begin{aligned} t &= e^z/C \\ w &= i\ln\frac{it+C'}{it-C'} \end{aligned} \tag{4.21}$$

with $z = x + iy$, $t = \eta + i\zeta$, and $w = u + iv$. Note that a circumference is mapped into the line segment $v = -v_o$ with $-\pi < u < \pi$ while the real axis (A-A') is mapped into the line segment $v = 0$ with $\pi > u > 0$. The steady-state isothermal lines of the structure in Fig. 4.5(b) are mapped into horizontal line segments in the w–plane whereas the steady-state heat flow lines are mapped into vertical line segments. Therefore, the thermal resistance is equal to four times the thermal resistance found between any pair of isothermal surfaces in

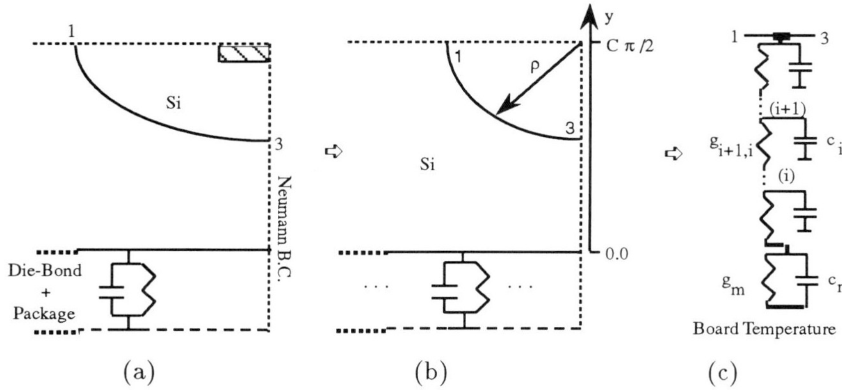

Figure 4.5 Thermal modeling for the material bounded by the *1-3* user-specified surface and the backside of the wafer. (a) Original structure, (b) reduced problem, and (c) equivalent *1D* model for the corresponding boundary condition.

the $z - plane$ given by

$$R_{i+1,i} = \frac{1}{g_{i+1,i}} = 4 \left[\frac{v_j - v_i}{2\pi\kappa v_o} \right] \tag{4.22}$$

where subscripts j and i denote the two isothermal surfaces. These labels can be uniquely associated with the intersections of the isothermal surfaces and the y–axis in the z–plane. This is the thermal resistance between any pair of nodes on *1D* grids of the *2D* structure. Thus, in terms of the geometric parameters of the original problem, one has

$$\begin{aligned} v_i &= \frac{1}{2} \ln \frac{C'^2 - 2C' \sin y_i / C + 1}{C'^2 + 2C' \sin y_i / C + 1}, \\ v_o &= \ln \rho / 2C, \\ C' &= \sqrt{1 - \rho^2 / C^2}. \end{aligned} \tag{4.23}$$

To determine appropriate models for the heat capacity of the material bounded by the *1–3* surface and the backside contact of the wafer, let us consider the thermal energy per unit length stored under steady-state conditions given by

$$U = \rho c_p \int_0^{C\frac{\pi}{2}} \int_{-\infty}^{+\infty} T(x,y)\, dx dy = \rho c_p \int_0^{-v_o} \int_{\pi}^{0} T(u,v) \left| \frac{\partial z}{\partial t} \right|^2 \left| \frac{\partial t}{\partial w} \right|^2 du dv. \tag{4.24}$$

For *1D* thermal modeling of the material bounded by the $1-3$ boundary and the backside contact of the wafer, this expression is used to compute the heat capacity associated with the material bounded by a pair of isothermals $i-\frac{1}{2}$ and $i+\frac{1}{2}$ that intersect the y–axis at locations $(y_i+y_{i-1})/2$ and $(y_i+y_{i+1})/2$, respectively. Under these conditions, we obtain

$$C_i = \frac{\pi}{4}\rho c_p C'^2 C^2 \left[|\coth v_i| - 1\right]\left[v_{i-\frac{1}{2}} - v_{i+\frac{1}{2}}\right]. \tag{4.25}$$

Finally, let us consider a method to properly determine a lumped model for the die-bond and the package material in the bottom part of the structure. If we deal only with a pure 2D simulation, this lumped model would correspond to a first-order RC network connected to each grid node in the bottom part of the simulation domain. The values of the RC elements would then correspond to the thermal resistance and the heat capacity per unit area. For the case of mixed 2D-1D thermal simulation, it becomes necessary to determine a characteristic dimension associated with the heat-flow line spreading to obtain a first-order RC model for the backside point of the wafer on the vertical principal axis of the defect. Finding this characteristic dimension is equivalent to determining the value of L_{eff}, depicted in Fig. 4.6(a), for which a fraction γ of the total Q flows through the backside contact. In the w–plane of Fig. 4.6(b), the region containing that fraction of the total energy is given by

$$\begin{aligned} u_s &= \pi - 2Arg(i + C') \\ u_1 &= \frac{\pi}{2}(1-\gamma). \end{aligned} \tag{4.26}$$

This can be immediately translated into the required value of L_{eff} by using the inverse transformation, i.e.,

$$L_{eff} = C \ \ln\left(C' \tan\left(\frac{\pi\gamma}{2}\right)\right). \tag{4.27}$$

Therefore, when the package material thickness beneath the die-bond is t_{pkg} and the corresponding thermal conductivity is κ_{pkg}, the lumped model for the die-bond and package material in the bottom of the structure is given by

$$\begin{aligned} g_m &= \frac{\kappa_{pkg} L_{eff}}{t_{pkg}} \\ C_m &= (\rho c_p)_{pkg} t_{pkg} L_{eff}. \end{aligned} \tag{4.28}$$

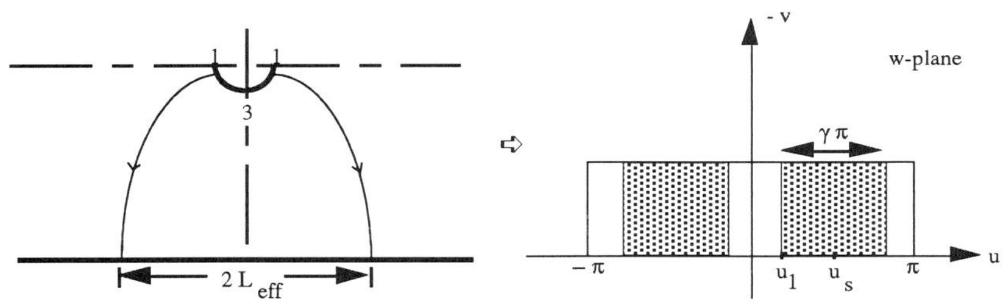

Figure 4.6 Schematic representation for L_{eff} at the backside of the wafer.

4.3 SIMULATION RESULTS FOR CERAMIC AND PLASTIC PACKAGES

Several simulations were run to determine the power vs. time-to-failure profile for different defect sizes, defect locations, and thicknesses of the passivation oxide in two different package systems, namely, plastic and ceramic. The corner values used in these experiments for each one of the variables under study are shown in Table 4.1, and the thermal parameters for the different materials involved are shown in Table 4.2. Note that the silicon thermal conductivity (lattice temperature model) and the density-specific-heat product for silicon are nonlinear functions of the lattice temperature. A schematic of the simulation domain used in this analysis together with the nomenclature used in Table 4.1 is shown in Fig. 4.2. The critical temperature was set to 1415 ^oC (silicon melting) for all of the **simulation** results presented in this chapter.

Table 4.1 Corner values used for two-level factorial experiments.

$\alpha[\mu m]$	$\beta[\mu m]$	$Y_j[\mu m]$	$t_{ox}[\mu m]$	Cera. Pack. Thick. $[mm]$	Plas. Pack. Thick. $[mm]$
0.5, 1.67	0.08, 0.4	0.38, 0.78	2.0, 5.0	2.5	3.4

The power profiles derived for a ceramic package and a 0.5 $\mu m \times 0.08$ μm defect site are shown in Fig. 4.7 for the four combinations of junction depth and oxide thickness (0 and 1 stand for the minimum and maximum values of a given parameter according to Table 4.1). Figure 4.7 indicates that the shape and the level of the power profile are not sensitive to the oxide thickness values, at least for values of t_f less than 10 ms. It appears that as the oxide thickness increases the power profile extends without changing its shape or location,

Table 4.2 Material physical parameters.

Material	$\rho[gm/cm^3]$	$c_p[J/gm-^o K]$	$\kappa[W/^o K - cm]$
Si	$0.897\,T^{0.1}$		$3104.15\,T^{-1.34}$
SiO_2	2.028		0.014
Air	1.16×10^{-3}	1.01	≈ 0.03
Alumina-96	3.72	0.77	0.18
Mold Compound	1.8	0.82	0.0106

which can in part be attributed to the low thermal conductivity of the oxide compared to that for the bulk silicon (see Table. 4.2). Simulation results in this figure also suggest that as the defect site moves deeper into silicon, the power profile shifts upward, as expected, due to the reduced thermal impedance seen by the heat source as it moves away from the oxide layer. Figure 4.7 also compares the simulation results with the predictions of the analytical model proposed by Dwyer et al. in the limiting case of an infinite parallelepiped. The silicon's thermal parameters are evaluated at two different temperatures — silicon melting (1415 ^oC) and aluminum melting (675 ^oC). It is apparent from the figure that the analytical model predictions depend greatly on the selection of temperature at which silicon's thermal parameters are evaluated. In this particular case, it is clear that the best fit of the analytical linear model (Dwyer et al.) to the nonlinear behaviour is obtained by evaluating the material parameters at about 700 ^oC. Due to the nonphysical underlying assumptions, such analytical models cannot be used reliably to investigate the impact of heat source location, the passivation oxide or package materials.

The same trends were observed for three other combinations of defect sizes. For each defect size, the power profiles fall into a region that is bounded by the cases corresponding to the combinations 00 and 11 of the $Y_j T_{ox}$ variables in Table 4.1. Figure 4.8 shows the bounding power profiles for the various combinations of defect sizes considered. In general, an increase of the defect size, while keeping the total power delivered to the structure constant, results in an increase of the time-to-failure, t_f, in the submicrosecond regime as shown in Fig. 4.8. For t_f longer than 0.1 μs, the spread of power profiles with respect to defect size is reduced. In some cases, the power profiles for different defect sizes eventually cross over each other depending on the values of the passivation oxide thickness and defect location, as demonstrated by the power profiles for defect site size cases 10 and 11 in Fig. 4.8. These trends are physically reasonable and agree well with analytical models that predict a direct proportionality between the temperature response and the power density [43].

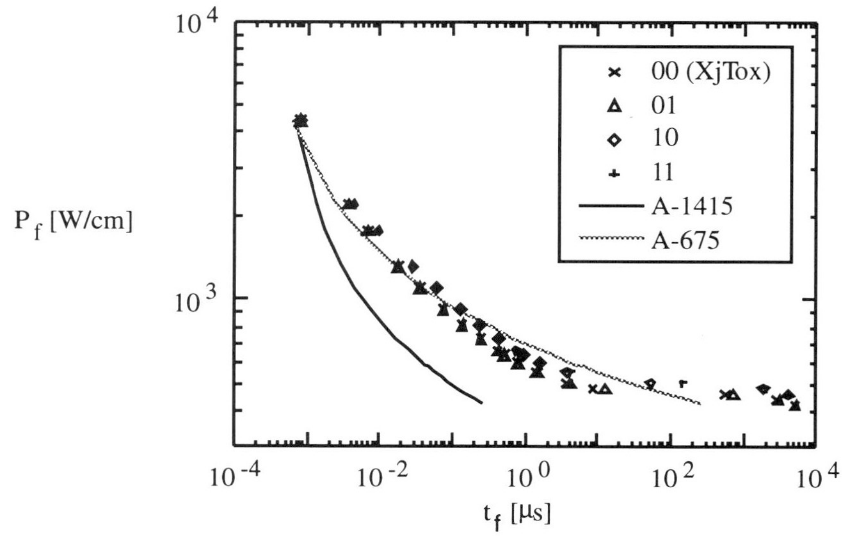

Figure 4.7 Mixed *2D-1D* iTSIM simulation results for a typical ceramic package and a 0.5 $\mu m \times 0.08\ \mu m$ defect site size. The analytical solutions using the model od Dwyer et al. (with silicon's thermal parameters evaluated at two different temperatures) is provided for comparison.

iTSIM's results for the typical plastic package case are summarized in Fig. 4.9, wherein the minimum and maximum defect size cases provide the lower and upper bounds of the power profiles. The results also indicate that the change of the oxide thickness simply shifts the times-to-failure without changing the power profile shape. Again, the power profile shifts upward as the defect location moves deeper into silicon. Figure 4.9 also compares the simulation results with the predictions of the analytical model of Dwyer et al. in the limiting case of an infinite parallelepiped assuming that the silicon's thermal parameters are evaluated at 675 ^oC. For short times-to-failure ($< 10\,\mu s$), this figure suggests that analytical models can be fitted to accurate nonlinear numerical results by properly selecting the temperature at which the thermal properties of the material are evaluated. However, for long times-to-failure ($> 100\,\mu s$), analytical models underestimate the power-to-failure with respect to the iTSIM's results. This can be attributed to the fact that analytical models assume an infinite medium and, therefore, do not account for heat sink properties of the package material.

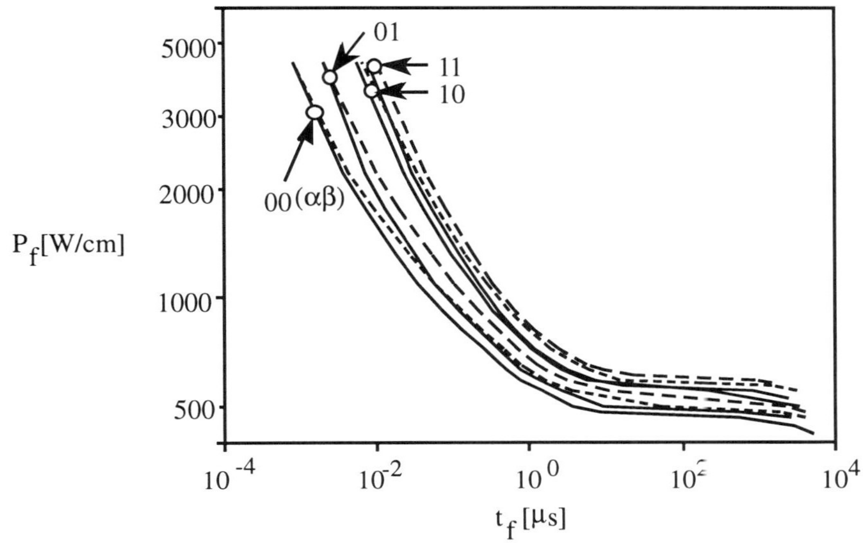

Figure 4.8 Bounding simulated power profiles for the various combinations of defect sizes considered here for a ceramic package.

It is important to note that the simulated failure power profiles presented in Figs 4.7, 4.8 and 4.9 lack the adiabatic regime $(1/t_f)$ for very short times. For the defect dimensions considered in the simulation examples, based on the model of Dwyer et al., one finds that the adiabatic regime of the power profiles is confined to times-to-failure less than 11.6 ns or 27 ns if silicon parameters are evaluated at 675 ^oC or 1415 ^oC, respectively. Our simulation results agree with the expected behaviour predicted by analytical models.

Among a given set of devices, the device which takes the largest power to cause failure for any given stress duration is said to have the best EOS/ESD performance. This means that its power profile would be an upper bound for all of the power profiles in the particular set of devices considered. The worst and best EOS/ESD performances measured in terms of the power profiles for the cases analyzed are shown in Fig. 4.10. For the same power level in a defect, the worst performance case would correspond to a minimum size defect (see Table 4.1) while the best performance case would correspond to the maximum size defect. It is observed that the package type and package material characteristics are not dominant factors in determining the profile shape for the cases analyzed. However, for times-to-failure greater than 10 ms, the profile

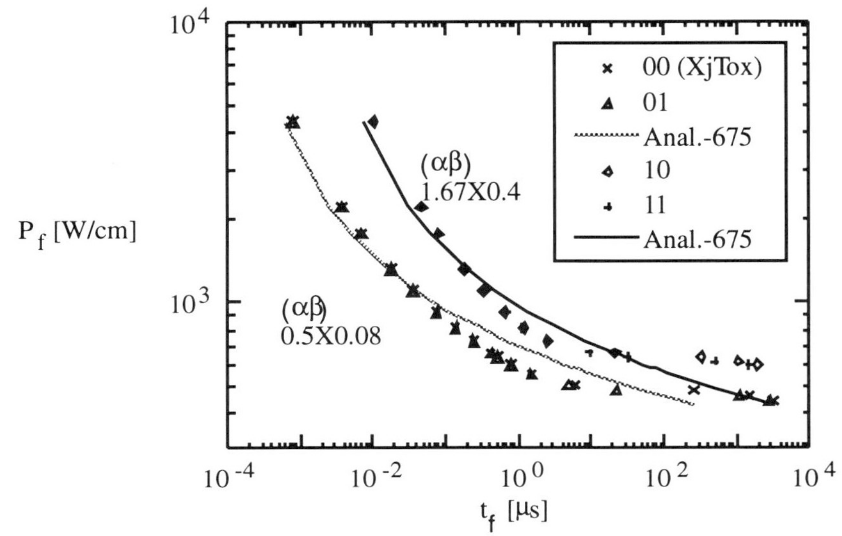

Figure 4.9 Typical power profiles for a plastic package, derived using iTSIM mixed *2D-1D* simulation. The analytical solution using the model of Dwyer et al. (silicon parameters evaluated at 675 ^oC) is provided for comparison.

shape may become increasingly dependent on the packaging material.

Assuming that the third dimension of the defect (heat generating volume) in a given device equals its width for all times-to-failure, one can compare the simulated power profiles with the experimental data [23, 24]. The experimentally derived power profiles are dimensionally reduced by dividing the measured power-to-failure by the device width. Figure 4.11 shows the comparison of these profiles for various device designs together with the *2D-1D* simulation results obtained on plastic packages. For t_f less than 5 μs, the simulated power profiles underestimate the time-to-failure for a given stress power level. In other words, for a given short time-to-failure, the stress power level required to reach failure is underestimated by the simulator, an effect that can be attributed to the 2D nature of the iTSIM simulation. In iTSIM, heat flow is constrained to two dimensions while in actual devices heat flow is primarily three-dimensional, thus reducing the thermal impedance seen by the source. This effect will also appear in the *2D* electrothermal simulation results presented in Chapter 6. Note also that the dimensionally reduced experimental power profiles, shown in Fig. 4.11, provide a different device ranking compared to those for the cor-

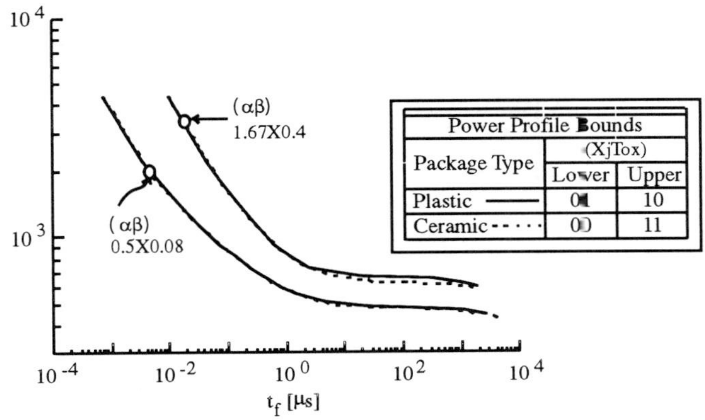

Figure 4.10 Comparison of the limiting power profiles for the plastic and ceramic packages as predicted by iTSIM.

responding *3D* measured profiles in Fig. 3.8. This implies that while the *3D* power profiles determine the relative EOS robustness for a given set of devices, the dimensionally reduced profiles indicate how efficiently, in terms of silicon area, a given layout provides EOS protection. Therefore, Fig. 4.11 suggests that room exists for area-efficient device hardening against EOS, at least for stress events longer than 10 μs.

4.4 SUMMARY

As a first step to the development of design guidelines for EOS/ESD hardening of ICs, the sensitivity of the power profiles to major thermal parameters of the die-package system was investigated. To this end, a nonlinear mixed *2D-1D* thermal simulator, iTSIM, has been developed. A proper set of boundary conditions has been introduced to achieve computational efficiency for large-scale EOS thermal problems.

Simulation results show that power profiles are determined mainly by the heat source size, its location with respect to the passivation oxide, and to a lesser extend the thermal characteristics of the package. As expected, our simulation results indicated that the EOS performance measured by the power profiles improves as the power density decreases (heat source size increases). Simulation

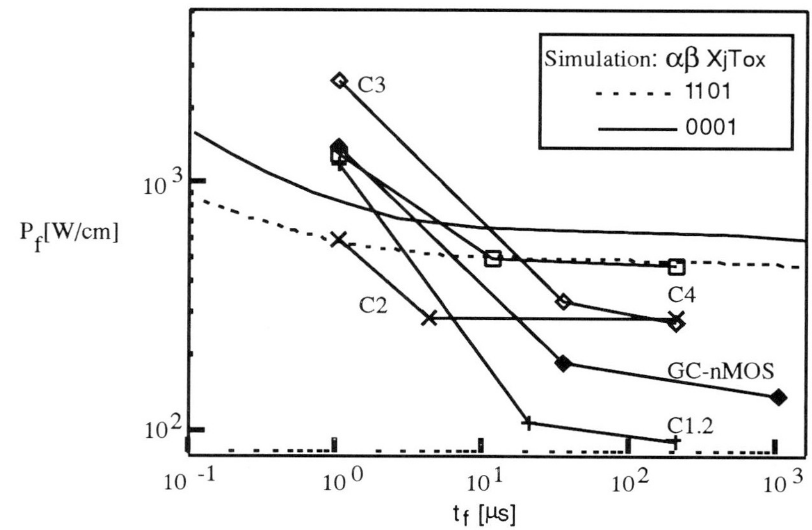

Figure 4.11 Comparison of the simulated power profiles with dimensionally reduced experimental data for the plastic package case.

results also reveal that the EOS performance improves as the heat source moves deeper into silicon, an effect caused by the reduced thermal impedance seen by the source as it moves away from the oxide layer. Simulation results also suggest that the packaging material does not dominate the profile shape for times-to-failure less than 10 *ms*. The curvature of the profile is not greatly affected by the thickness of the passivation oxide; for ceramic packages, an increase of the oxide thickness causes the profiles to extend without changing its shape or location. For short times-to-failure, though, the $2D$ nature of iTSIM simulation underestimates the power required to cause failure compared with experimental data, the presented simulations results have provided a physically reasonable quantitative assessment of geometrical details and material nonlinearities.

It is worthwhile to note that the experimental power profiles, when dimensionally reduced by dividing the power P_f by the device width as shown in Fig. 4.11, are not on the same order as the actual *3D* power profiles reported in [23, 24]. This means that while the 3D power profile determines the relative EOS robustness among a given set of devices, the dimensionally reduced profiles can be used to measure the EOS protection efficiency of a given layout in terms of the silicon area.

To gain more physical insight into the phenomena leading to device failure during EOS/ESD events, electrothermal analysis is required. The boundary conditions presented here should be implemented in $2D$ electrothermal device simulators with the goal of studying long lasting EOS phenomena with appropriate accuracy while preserving adequate computational efficiency.

5

ITSIM: A NONLINEAR $2D - 1D$ THERMAL SIMULATOR

5.1 INTRODUCTION

iTSIM[1] is a nonlinear thermal simulator initially designed to study the power profile sensitivity to major IC thermal parameters. The program is contained in 15 source files written in the *C* language and is provided with a *makefile* to compile it and install the executable binary that runs under a UNIX operating system.

iTSIM solves the nonlinear heat diffusion equation using the finite differences scheme. ITSIM can perform both steady-state and transient analysis in either *1D* or *2D*. Temporal discretization in iTSIM is done using either the backward Euler or the trapezoidal rule schemes. The resulting nonlinear discrete system of equations is solved using Newton's method. The sparse matrix solver in [57] is used to solve the linearized system.

iTSIM can either perform a true *2D* simulation or a mixed *2D-1D* simulation that is suitable for large-scale thermal problems. iTSIM uses three types of boundary conditions, namely, Dirichlet BCs (package heat sinks), Neumann BCs (adiabatic), and the special set of boundary conditions introduced in Chapter 4 that enable efficient simulation of large-scale thermal problems.

The current version of iTSIM was aimed at the study of the sensitivity of IC EOS power profiles with respect to major IC thermal parameters. iTSIM exploits the symmetry of the *2D* EOS thermal problem with respect to the principal axis of the heat source perpendicular to the material interface. Along this axis, iTSIM always uses adiabatic boundary conditions.

[1] The source code can be obtained upon request from the University of Illinois.

5.2 RUNNING THE PROGRAM

To run iTSIM, the user requires an input file that describes the simulation domain, the material parameters, and the desired analysis and output format. The syntax of that input file is described in the next section. The generic command line syntax of iTSIM is

iTSIM_Sp [-b] [-v] [-t] inputfile [outfile].

The output file name is optional; if not given, the output will be directed to the standard output. If no flags are provided in the command line, the simulator will use the trapezoidal rule integration method for transient analysis with variable time step. The flag *[-b]* would instruct the simulator to use the backward Euler integration method with the user-provided fixed time step for transient analysis. The flag *[-v]* would instruct the simulator to use the backward Euler integration method with a variable time step for transient analysis. The flag *[-t]* would force the simulator to use the trapezoidal rule integration method with a fixed time step for transient analysis.

5.3 INPUT FILE

The syntax of the input file is similar to that for a typical PISCES input deck. Except for the first line that is always taken as a title card, no specific order is assumed for the control cards. Any line that starts with $*$ is treated as a comment line. iTSIM is not case-sensitive since the input parser capitalizes the input lines before processing them. In the following subsections, the syntax and description of valid input lines for iTSIM are given. A sample input file and simulation result can be found in the next section.

5.3.1 Grid Specification

Grid specification in iTSIM is done with the following statements:

xyscale 1.0e-4
GRIDY N1 YLOC1 SF1 $\cdots$ Nn YLOCn SFn
GRIDX N1 XLOC1 SF1 $\cdots$ Nn XLOCn SFn
MIX2D1D NYB LYE

Examples:

xyscale 1.0e-4
gridy 1 -4.0 1.0 10 0.0 0.95 14 0.4 1.0 26 6.0 1.1
gridy 39 12.0 0.9 40 12.2 1.0 70 254.0 1.2
gridx 1 0.0 1.0 5 0.4 1.0 15 6.0 1.2 29 11.9 0.87 30 12.0 1.0
mix2d1d 18 40

The optional SCALEXY card can be used to modify the internal scaling factor whose default value is *1.0E-4*. This factor is used to convert the grid line location values to centimiters. GRIDY and GRIDX specify the Y and X grids, respectively. Each set of three numbers in these cards is comprised of a grid line number N, a line location LOC, and a geometric exponent SF. There should be at least two of these sets per grid card in an acceptable input deck. The first grid line should always be numbered as 1. The location determines the position of the particular grid line along the corresponding axis and should be given in microns. The geometric exponent should always be positive and determines how the grid spacing varies across the interval defined by the location of the grid line in the current and previous date set of numbers. A value of 1 for this parameter causes the program to produce a uniform grid. If the geometric exponent is less than one, the spacing between grid lines decreases geometrically as the lines approach to the location specified in the set that contains the geometric exponent. If the geometric exponent is less than one, the spacing between grid lines grows geometrically as the lines get close to the location specified in the set that contains the geometric exponent.

An X–or Y–grid specification may be contained in more than one line. In this case, each line should contain complete sets of parameters. Any other line after the first is treated as a continuation of the grid specification, and the sequence in line numbers and locations should be maintained.

At least one GRIDY statement should be specified in the input deck. For the Y grid, there must always be a grid line located at y = 0.0 that corresponds to the interface between the two materials. For *1D* simulation cases, no GRIDX line should appear in the input deck.

The MIX2D1D card is used to force iTSIM to employ the mixed *2D-1D* simulation approach. In this case, the *2D* grid is terminated in a boundary defined by NYB and NYE as follows. Beginning at the grid node defined by the intersection of the Y-grid line NYB and the last X-grid line specified in a GRIDX card, the algorithm begins to remove grid nodes by terminating pairs of Y and X grid lines in staircase fashion. If the X– and Y– grid specifications (spacings)

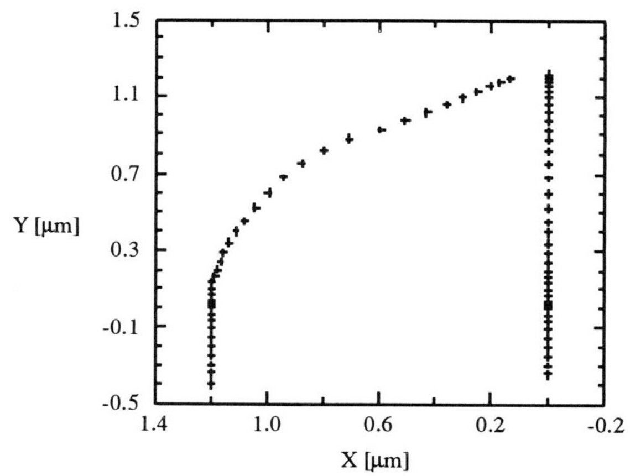

Figure 5.1 The mixed *2D-1D* simulation boundary generated by iTSIM from the grid specification cards shown in the example above. Only the *2D* portion of the simulation domain is shown here.

are well-defined, this should result in a *2D* simulation domain terminated by a boundary that is approximately an arc of circumference (the *1-3* boundary in Fig. 4.2). Notice that beyond the location of the line NYE, the grid becomes *1D* as defined by the GRIDY card. Figure 5.1 shows the boundary that iTSIM would generate given the grid specification cards in the example above. In this example, according to the data in the MIX2D1D and GRIDY cards, the grid should become one dimensional beyond the Y-grid line number 40, which is located at 12.2 μm below the Si-SiO_2 interface.

The SCALEXY card can be used to modify the internal scaling factor whose default is $1.0E-4$. This factor is used to convert the grid line location values to centimeters.

5.3.2 Material thermal parameter specification

The material specification to iTSIM is done using the following mandatory cards

MAT0 KO KTE PCPO PCPTE

MAT1 KO KTE PCPO PCPTE

Examples (silicon dioxide and silicon):

mat0 0.014 0.0 2.028 0.0

mat1 3104.14 -1.33 0.897 0.1

MAT0 and MAT1 define the thermal parameters associated with the materials above ($y < 0$) and below ($y > 0$) the interface, respectively. The four parameters in each card are used to determine the temperature-dependent thermal conductivity and specific heat of each material according to

$$\kappa(T) = KO \times T^{KTE}, \; [W/cm -^{o} K]$$
$$(\rho c_p)(T) = PCPO \times T^{PCPTE}, \; [J/cm^3 -^{o} K].$$

5.3.3 Package thermal boundary conditions

The thermal properties of the package, determining the boundary conditions at the backside of the wafer and at the oxide-package interface are specified to iTSIM by using the following mandatory card:

BOUND RA CA RM CM

Example (ceramic package):

bound 416.7 0.0 0.694 0.36

BOUND specifies to iTSIM the die's bottom and top thermal boundary conditions. Two pairs of data are expected in this card. Each data pair represents the thermal resistance $[^oC - cm^2/W]$ and thermal capacity $[J/^oC - cm^2]$ per unit area of the material in contact with the die at the corresponding surface. iTSIM associates the first data pair in this card with the interface determined by the Y-grid line number 1 (usually oxide-package interface). The second data pair is associated with the interface at the last Y-grid line (die's backside package interface) to material 0, while the second is for material 1.

5.3.4 Heat source

The general format of this mandatory card is

SOURCE VALUE PW NYMIN NYMAX NXMIN NXMAX

Example:

source 3.0e+12 1.0e-3 11 12 30 30

The SOURCE card specifies to iTSIM the region where a heat pulse is to be applied. The card requires six parameters. The first parameter is the value of the heat source per grid point in the generation region $[W/cm^3]$. The second parameter is the pulse width in seconds. The other four parameters are the Y and X grid line numbers that define the heat source region (actually the left half since iTSIM forces the user to exploit the symmetry of the problem). Necessary conditions for the source region to be considered well-defined are that $NYMIM \leq NYMAX$, $NXMIN \leq NXMAX$, and NXMAX should be equal to the maximum grid line number specified in the GRIDX card. For *1D* simulations no GRIDX card should be present and NXMAX and NXMIN should be set to 1 in the SOURCE card.

5.3.5 Analysis specification

iTSIM can perform either steady-state (dc) thermal analysis or transient thermal analysis. By default, iTSIM performs an operating point analysis if the input deck does not specify either DC or TRAN.

The general format of a *DC* card is

DC HINIT HSTOP HSTEP

where the heat source value is swept from HINIT to HSTOP with steps of value HSTEP.

.TRAN TSTEP TSTOP

In this case iTSIM will perform transient simulation starting at time 0 and stopping at TSTOP. If a fixed time step is used, TSTEP is the simulation time step. For a variable time step, TSTEP is used to determine the minimum and maximum time steps using the corresponding default or user-provided time-step factors.

5.3.6 Output format

One type of output request specification is

PRINT GRID .

Whenever an input file contains this card, iTSIM will compute the grid, print a summary of it, and exit without performing any simulation. This card is useful when the user is preparing the grid specification particularly when using mixed *2D-1D*, since it is important to have an appropriate shape of the $1-3$ boundary 4.2. A visualization of the *2D* grid boundaries as shown in Fig. 5.1 can be obtained using the plotting package XGRAPH [49].

For operating point analysis, iTSIM prints the temperatures at each grid node in fixed format. In this case no *print* card should be specified.

If iTSIM is requested to perform a *DC* or *TRAN* analysis, a *print* card should be included with the following general format:

PRINT TYPE [NX1 NY1 ··· NXN NYN]

where type is either *DC* or *TRAN*, and the NXi and NYi determine the grid nodes whose temperature should be printed. For transient analysis, it is always advisable to monitor at least the temperature at the center of the heat source since the time step control routine monitors only the rate of change of nodes in the *print* card. When using mixed *2D-1D* simulation, the user provided NXi in the *print* card should be **adjusted** to reflect the X-grid line termination.

An input deck may have more than one *.PRINT* card, provided each card has complete pairs of data and that the total number of requested outputs is not greater than 10. (This can be increased by changing the value of MNOV in the header file.)

The default output format in iTSIM is tabular. The default can be overridden if the user includes an *XGRAPH* card in the input deck. In this case, iTSIM will generate a file that after minor modifications can easily be plotted using the plotting package XGRAPH [49].

5.3.7 Options and debugging

The generic option card format is *.OPTIONS [ONAME = VAL]*. The following options are available:

Name	Description	Default
ACCFLAG	enables accuracy checks in pure *2D* grids only, results are written to [outfile]x.tmp	FALSE
BTEMP	board temperature value	$27[oC]$
DX	relative tolerance for convergence	1e-4
ITLIM	set the iteration limit	15
LTE	local truncation error to control the time step	1e-3
MINTSF	factor used to determine the minimum time step from TSTEP	1e-3
MAXTSF	factor used to determine the maximum time step from TSTEP	100.0
R	factor used in determining L_{eff} for the backside BCs (cf. Fig. 4.6)	0.99
MINDTDT	if during transient simulation the rate of change of any monitored temperature is $< MINDTDT$, the simulation stops	5e4 $[^oC/s]$
TCRIT	if any node in the structure reaches $TCRIT$ the simulation stops	1415 $[^oC]$
ULSBC	forces iTSIM to use nonadiabatic BCs at each grid node on the $1-2$ boundary (cf. Fig. 4.2)	TRUE
ULSTAVC	factor used to determine the linearization temperature for the material properties of grid nodes on the $1-2$ boundary (cf. Sec. 4.2.1)	0.9

5.4 AN EXAMPLE

A typical partial printout generated by iTSIM is presented is shown in Fig 5.2. The case under analysis is an IC with a $4.0\,\mu m$ passivation oxide layer on top of a $254\,\mu m$ silicon die in a ceramic package. The heat source is assumed to be $0.1 \times 0.2\,\mu m^2$ located $0.1\,\mu m$ below the Si-SiO_2 interface; mixed *2D-1D simulation* is used. In this example, the failure condition was met at $89.1\,ns$ as can be seen from the printout. iTSIM always reports the grid coordinates of the node that satisfied the failure condition.

```
<<<*** iTSIM (1.2) Carlos H. Diaz.   Thu Mar 18 10:40:38 1993 ***>>>
>>>> 2D case. 4.0um tox, 254um wt, xj 0.1um, defect 0.1X0.2, Ceramic
*
gridy 1 -4.0 1.0  10 0.0 0.95  14 0.4 1.0  26 6.0 1.1
gridy 39 12.0 0.9 40 12.2 1.0 70 254.0 1.2
mix2d1d 18 40
*
gridx 1 0.0 1.0 5 0.4 1.0 15 6.0 1.2 29 11.9 0.87 30 12.0 1.0
*
mat0  0.014   0.0   2.028   0.0
* silicon temperature dependent thermal parameters
mat1  3104.14  -1.33   0.897 0.1
*
bound 416.7 0.0 0.694 0.36
source 3.0e+12 1.0e-3 11 12 30 30
*tran 1.0e-9 1.0e-3
tran 1.0e-9 89.1e-9
print tran 30 1 30 10 30 11 1 40 1 70
options lte = 2.0e-3 dx = 1.0e-3  ULSBC = 1
end
*** Failure at (nx = 30, ny = 11) 8.910e-08 [s], 1.415e+03 oC.
<<<*** iTSIM (1.2) Carlos H. Diaz.   Thu Mar 18 10:46:57 1993 ***>>>
>>>> 2D case. 4.0um tox, 254um wt, xj 0.1um, defect 0.1X0.2, Ceramic
Peak Power delivered:  3.000e+02 [W/cm]
Time [us]   T[30] oC    T[300] oC   T[330] oC   T[918] oC   T[948] oC
0.000e+00   2.700e+01   2.700e+01   2.700e+01   2.700e+01   2.700e+01
1.000e-06   2.700e+01   2.757e+01   2.757e+01   2.700e+01   2.700e+01
2.000e-06   2.700e+01   2.869e+01   2.870e+01   2.700e+01   2.700e+01
3.000e-06   2.700e+01   2.980e+01   2.981e+01   2.700e+01   2.700e+01
4.000e-06   2.700e+01   3.089e+01   3.090e+01   2.700e+01   2.700e+01
8.108e-06   2.700e+01   3.521e+01   3.523e+01   2.700e+01   2.700e+01
1.193e-05   2.700e+01   3.900e+01   3.903e+01   2.700e+01   2.700e+01
.
.
.
8.228e-02   2.700e+01   1.347e+03   1.374e+03   2.706e+01   2.700e+01
8.507e-02   2.700e+01   1.363e+03   1.391e+03   2.708e+01   2.700e+01
8.796e-02   2.700e+01   1.379e+03   1.408e+03   2.709e+01   2.700e+01
8.910e-02   2.700e+01   1.386e+03   1.415e+03   2.710e+01   2.700e+01
Interpolated values for failure condition:
8.910e-02   2.700e+01   1.386e+03   1.415e+03   2.710e+01   2.700e+01
Total run time = 365.730 CPU sec.
```

Figure 5.2 Typical print-out generated by iTSIM for an IC in a ceramic package.

6

2*D*-ELECTROTHERMAL ANALYSIS OF DEVICE FAILURE IN ADVANCED MOS PROCESSES

The most common protection schemes used in MOS ICs rely on the parasitic bipolar transistor associated with the nMOS devices. A basic MOS protection network includes an nMOS device whose drain is connected to the pin to be protected with the source tied to ground. The protection level or failure threshold can be set by varying the nMOS device width, provided uniform conduction is achieved. Under stress conditions, the dominant current conduction path between the protected pin and ground involves the parasitic bipolar transistor associated with the nMOS device. This parasitic bipolar transistor operates in the snapback region when the pin is stressed positive with respect to ground. Typical breakdown characteristics of a grounded-gate nMOS device are shown for reference in Fig 6.1. Generally, the nMOS protection device will fail under ESD/EOS as a result of the second breakdown which causes the current localization and the eventual device failure (thermal runaway) [87, 5]. Thermal runaway is a positive feedback process in a circuit component produced by the coupling of the device self-heating and the temperature-dependent electrical parameters under stress conditions.

This chapter presents the device-level electrothermal analysis performed to explain the experimental EOS characterization results presented in Section 3.3 for the MOS gate-coupled ESD protection structures.

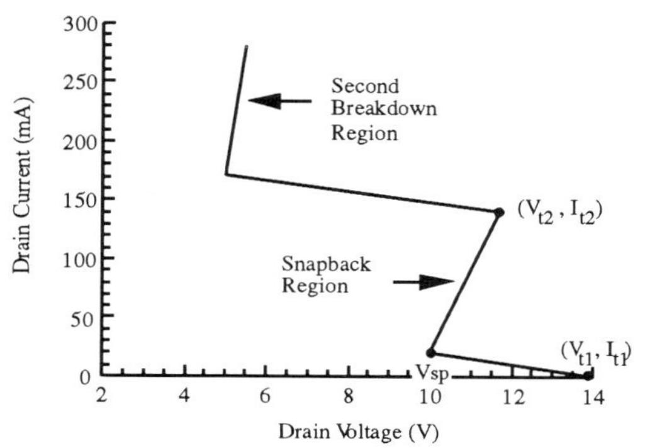

(a)

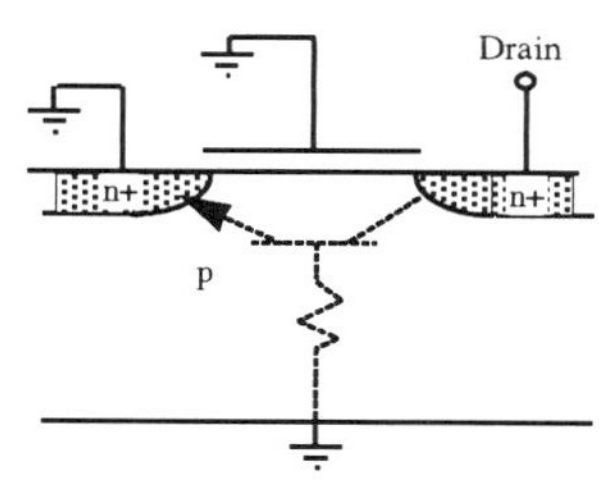

(b)

Figure 6.1 (a) Typical *I-V* characteristics for an nMOS device showing snapback and second breakdown regions. (b) Schematic cross-section of the nMOS device indicating its parasitic *npn* transistor.

6.1 DEVICE LEVEL ELECTROTHERMAL SIMULATION

As described in Section 3.3, nMOS gate-coupled protection devices with varying widths and drain and source contact-to-gate spacings were fabricated and characterized for ESD and EOS to determine the impact of major layout parameters on the ESD/EOS performance. Electrothermal simulations were done using the *2D* device simulator TMA-MEDICI [97], which was run with the circuit analysis and the lattice temperature modules in place. TMA-MEDICI uses Adler's formulation [2, 90] for heat generation in nondegenerate materials. This formulation accounts for Joule heating and lattice energy loss/gain due to generation/recombination.

The simulation results presented in this section were obtained using a step voltage source with 50Ω series resistance connected to the drain of the nMOS device under stress. The gate of the device was connected to a resistor (replacing the FOD in Fig. 3.17(a))[1]; this resistor and the source of the device were connected to ground. The width of the *2D* simulation domain, similar to the one shown in Fig. 3.15, was determined by the drain and source contact-to-gate spacings, the contact size, and the channel length. The depth of the simulation domain

[1] This approach is valid since the characteristic failure times of the stress levels considered here are much longer than the transistor turn-on time. The simulation results presented here were obtained using a 100 Ω gate resistor.

was set in all cases to 10 μm. The grid node count was kept below 1800 for the cases considered as a compromise between accuracy and simulation speed — the average run time of a typical EOS transient simulation is about 5 hrs in a Sparc IPX workstation. A lumped resistance connected to the bottom side of the structure was used to model the current flow to the backside contact of the wafer. The value of this resistance was computed from [88, 60]

$$R_{sub} = \frac{\rho_s}{2\tan(\theta)} \ln\left(1 + \frac{2t\tan(\theta)}{w}\right)$$

where ρ_s is silicon's resistivity, θ is the substrate current spreading angle, t is approximately equal to the die thickness, and w is the simulation domain width. For a 12.6 μm wide simulation domain and a die thickness of 250 μm, the value of the substrate resistance is approximately 185.7 $\Omega - \mu m$ in this process. Adiabatic boundary conditions were used at all boundaries except the bottom side of the simulation domain where heat removal was allowed by connecting a lumped linear thermal resistance and capacitance to model the heat conduction to the die's backside which was assumed to be held at room temperature. The values of these elements were calculated according to the results presented in Chapter 4: Eqs. (4.22) and (4.25). For a die thickness of 250 μm and a 10 μm deep simulation domain, the values of the lumped thermal resistance and capacitance were estimated as 1.067×10^4 ^oC-$\mu m/W$ and 4.24×10^{-9} $J/\mu m$-^oC, respectively. For the thermal boundary conditions, silicon's thermal parameters were evaluated at an empirically chosen temperature of 600 ^oK since TMA-MEDICI allows only linear thermal boundary conditions.

Figure 6.2 shows the simulated response of the pad (drain-source) voltage to a 1.85 A current pulse with 2 ns rise time. The device was 500 μm wide with drain and source contact-to-gate spacings (DCGS, SCGS) of 6 μm. As seen in Fig. 6.2, the device goes into breakdown at about 11 V and then snaps back to approximately 6.2 V. As the current is ramped up, pad voltage builds up mainly as a result of the increased voltage drop across the drain/source contact resistances. After 2 ns, the device heating effect becomes noticeable in the drain voltage waveform. Initially, and up to about 21 μs, the increase of the pad voltage is a result of the device heating that causes a reduction in the current gain of the parasitic bipolar transistor[2] and a reduction of the impact ionization rates [18, 90]. From Fig. 6.2(b), the device continues heating up but soon after 21 μs the pad voltage begins to decrease because the device conduction is dominated by the thermal carrier generation that has become high enough to offset the effects of the current-gain degradation and the reduction of

[2] The current gain of the parasitic bipolar transistor is proportional to the mobility of the minority carriers in the channel region [93, 95].

the impact ionization rates. This negative differential resistance phenomenon marks the onset of second breakdown for this device. The slight bowing seen in the current waveform near the time for second breakdown is due to the series resistance of the stress source. The waveforms shown here are representative of the results obtained for various stress level conditions on different devices.

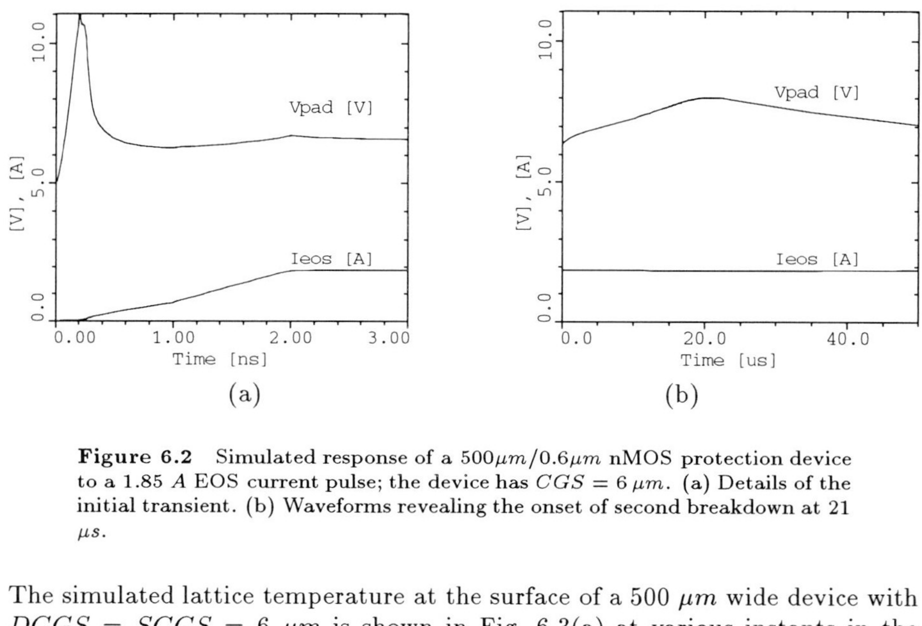

Figure 6.2 Simulated response of a $500\mu m/0.6\mu m$ nMOS protection device to a 1.85 A EOS current pulse; the device has $CGS = 6\ \mu m$. (a) Details of the initial transient. (b) Waveforms revealing the onset of second breakdown at 21 μs.

The simulated lattice temperature at the surface of a 500 μm wide device with $DCGS = SCGS = 6\ \mu m$ is shown in Fig. 6.3(a) at various instants in the stress event. The simulation results correspond to a 1.85 A current pulse that led to second breakdown after 21 μs. As expected, this figure shows that the location of the peak lattice temperature coincides with the LDD drain junction. The figure also reveals the steepness of the lattice temperature gradients in the neighborhood of the LDD drain junction ($\sim 500\ ^oK/\mu m$). It can also be seen that the lattice temperature at the surface increases monotonically with time, at least up to the time for second breakdown. Although not shown here, this is also true for device lattice temperature all across the device. Figure 6.3(b) is a plot of the electric potential along a line parallel to the Si-SiO_2 interface for the same device. Figure 6.3(b) reveals that the voltage across the drain depletion region near the gate edge increases as the device heats up reaching a maximum at the time of second breakdown. After this, the drain voltage drastically decreases with respect to its maximum value. The figure also suggests that the effect of lattice heating on the voltage drop across the drain and source diffusions is not important compared to the temperature effects seen on the depletion region. Therefore, thermal runaway in these devices is due mainly to second breakdown of the parasitic *npn* transistor. The simulation results shown in Fig. 6.3 are also representative of the results for other stress levels and device layout parameters.

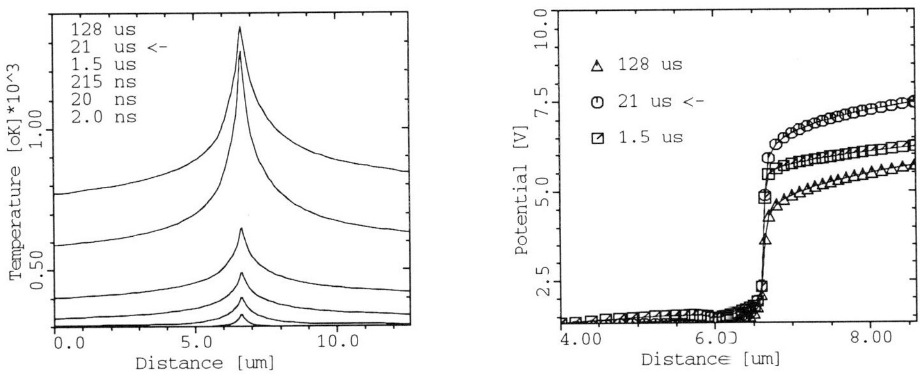

(a) (b)

Figure 6.3 A 500 μm / 0.6 μm nMOS protection device ($DCGS = SCGS = 6\ \mu m$) at various instants during a 1.85 A current pulse. The abscissa for the gate edge of the drain region is 6.6 μm (the distance is measured from the center of the source contact to the center of the drain contact). Plots of (a) the lattice temperature and (b) electric potential, along the Si-SiO_2 interface.

Figure 6.4 shows the 5% current density flow lines at the time of second breakdown for a 500 μm wide device with $DCGS = SCGS = 6\ \mu m$. As expected, the current density is higher near the surface and decreases away from it. As long as the device operates in the snapback region, the current distribution in this figure is also representative of the device conduction for the various stress levels and layout parameters considered here.

Transient electrothermal simulations to characterize the EOS performance of nMOS protection devices were run in two steps for each device design. A transient simulation was run to drive the device into the snapback region, for example, by ramping up the stress voltage source from 0 V to 40 V in 1 ns. The solution at the end of this simulation was saved. Then, several transient simulations were run for the corresponding levels of stress using the saved solutions as the initial guess, for example, by ramping up the stress voltage source

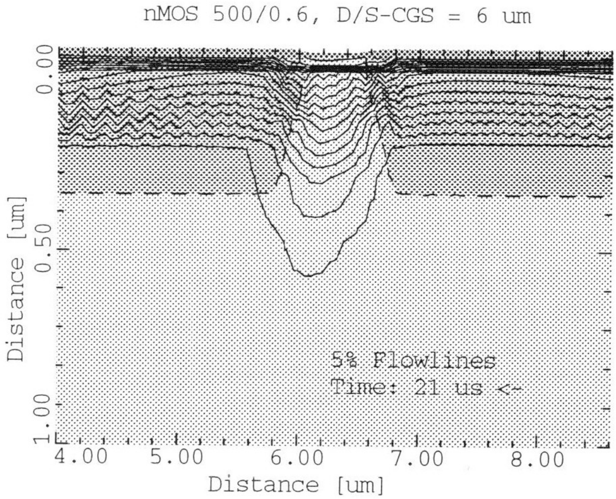

Figure 6.4 Current flow lines at the time of second breakdown for a 500 μm / 0.6 μm nMOS protection device having $DCGS = SCGS = 6\ \mu m$. The device is being stressed with a 1.85 A EOS pulse.

from 40 V to the desired stress level with a rise time of 1 ns. This approach was taken to speed up the simulation process, which is valid as long as the total rise time (2 ns in this case) is negligible with respect to the time-to-failure. The simulated transient drain voltage and drain current waveforms, similar to those shown in Fig. 6.2, can be translated into compact I-V plots by plotting the current waveform versus the voltage waveform. Figure 6.5 shows such I-V plots for devices with different values of the CGS, and subjected to different levels of stress current. The bifurcation of the traces is a result of the simulation approach just outlined. This plot indicates that increasing the contact-to-gate spacing results in increased snapback voltage and on-resistance. Furthermore, it suggests that by setting the value of the SCGS to the minimum allowed by the particular technology can lead to substantial savings of device power dissipation.

Two different device failure conditions were used for the analysis of simulation results. The time-to-failure was defined as either the time for the peak device temperature to approach the silicon melting point, or the time-to-second breakdown, which ever happens first. For the devices considered here, the simulation results indicated that very energetic stress pulses could cause the device to fail due to melting while, for other stress conditions, the device failure could be due to current localization resulting from second breakdown phenomena. Fig-

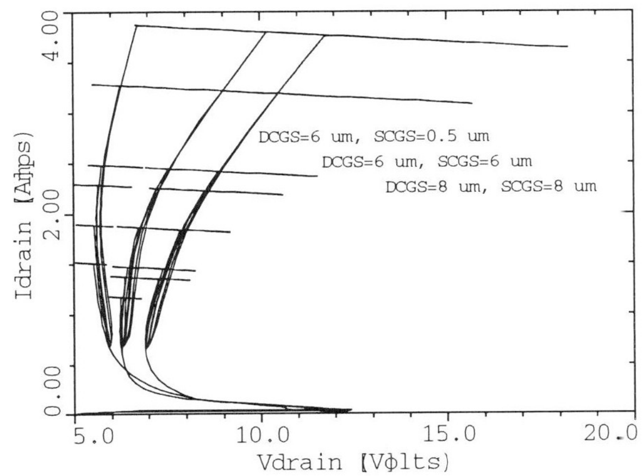

Figure 6.5 The simulated drain voltage and current waveforms at various levels of the stress current are shown as *I-V* curves. The results are given for three 500 μm / 0.6 μm nMOS devices having different contact-to-gate spacings.

ure 6.6 is a plot of the lattice temperature at the Si-SiO_2 interface at the time of second breakdown for a 500 μm wide device with 6 μm CGS. This plot shows how the steepness of the temperature gradient decreases as the stress current decrease; a result that is representative of the findings for other devices with different CGS values. Note that the temperature plots in Fig. 6.6 qualitatively explain the failure analysis results shown in Fig. 3.21; short-duration high-level stress events induce high peak temperatures near the drain that could lead to poly-gate filament formation. On the other hand, long-lasting low-level stress events could cause a larger portion of the device to be involved in the failure process.

6.2 COMPARISON OF EXPERIMENTAL AND *2D* ELECTROTHERMAL RESULTS

Various transient electrothermal simulations, similar to the one described in the previous section, were run for different levels of stress current and layout parameters to obtain simulated power and current profiles. Figure 6.7 shows

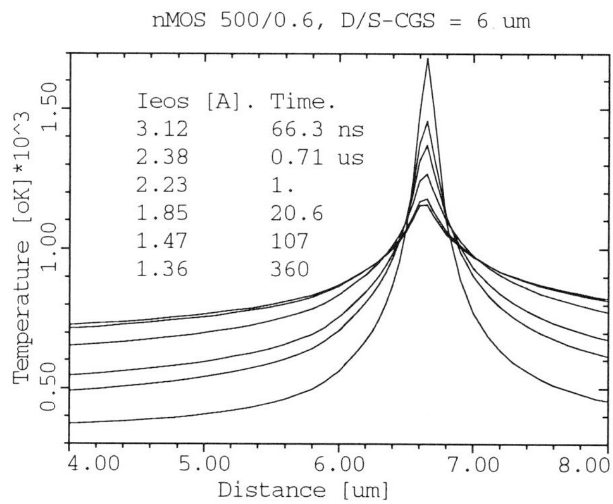

Figure 6.6 Lattice temperature along the Si-SiO_2 interface at the time of second breakdown for different levels of the stress current. The device is 500 μm wide with CGS = 6 μm. The peak temperature decreases, and the device heating tends to be more uniform as the level of the stress current decreases (the time for second breakdown increases).

the experimental and simulated power profiles for 500 μm wide devices having DCGS=SCGS. The figure also includes the simulated power profile for a 500 μm wide device having minimum SCGS. The figure indicates that simulation results tend to overestimate the power required to cause failure for pulse widths longer than 20 μs. This can be attributed to the lumped thermal model used to model the heat conduction from the bottom side of the simulation domain down to the backside of the die. For high power pulses, the simulation results underestimate the power required to cause failure, an effect of the *2D* nature of the simulation compared to the actual $3D$ heat flow (in a *2D* heat flow, the effective thermal impedance is higher than in a $3D$ heat flow, an effect also seen in the mixed $2D - 1D$ iTSIM simulation results presented in Chapter 4). The simulation results correctly predict the trends observed in experimental data, and can still be used to provide qualitative explanation of the phenomena leading to device failure as will be described below.

Figure 6.8 is a plot of the value of the stress current vs. the simulated time-to-failure for a 500 μm wide device. Clearly, this figure indicates that the onset of second breakdown depends on the level of the stress current. The results also suggest that only a marginal improvement in the failure current level can be

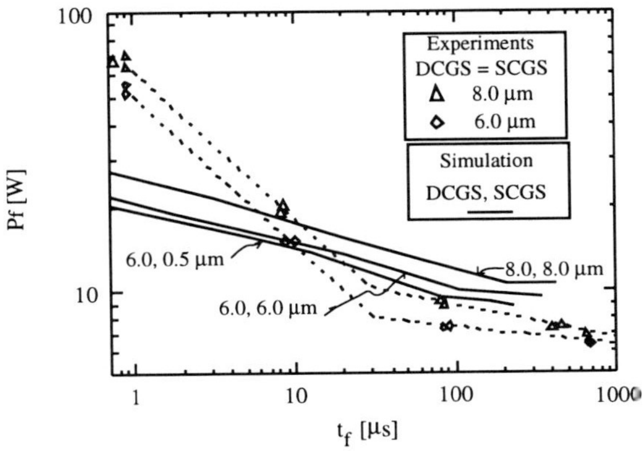

Figure 6.7 Comparison of experimental and simulated power profiles for 500 μm wide devices having DCGS=SCGS. Also shown is the simulated power profile for a 500 μm wide device with minimum SCGS.

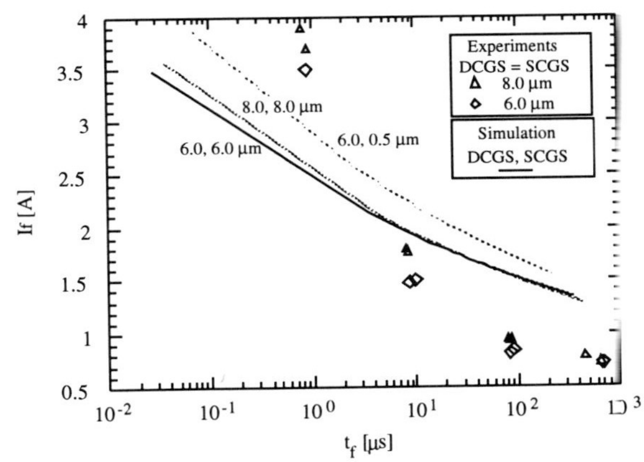

Figure 6.8 Experimental and simulated current profiles for 500 μm wide devices with different contact-to-gate spacings (CGS).

obtained by uniformly increasing the drain and source contact-to-gate spacings and that the EOS performance improvement measured by the corresponding power profiles can be attributed mainly to the increased power dissipation of the drain/source diffusions. However, if the SCGS is kept to a minimum while maintaining the same source area, Fig. 6.8 does predict a substantial improvement of the failure current level, thus, a higher failure threshold can be achieved. This result has been used in a design guideline for source contact placement [21].

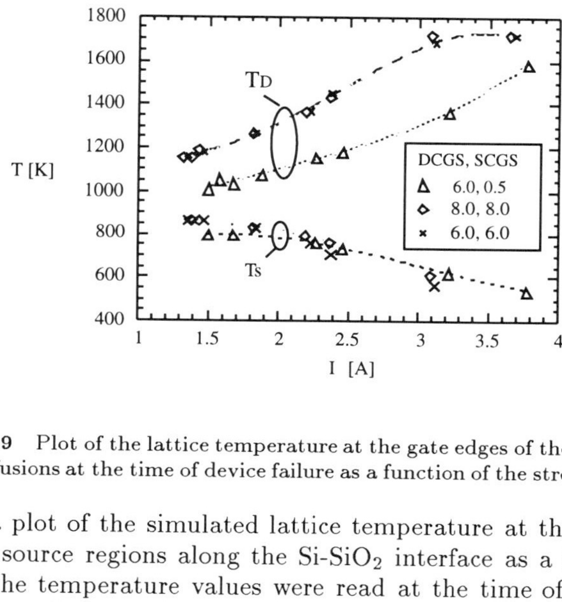

Figure 6.9 Plot of the lattice temperature at the gate edges of the drain and source diffusions at the time of device failure as a function of the stress current.

Figure 6.9 is a plot of the simulated lattice temperature at the gate edges of the drain and source regions along the Si-SiO_2 interface as a function of the stress level. The temperature values were read at the time of device failure. This figure shows that, as the stress level decreases, the temperature in the drain edge decreases due to the reduced heat generation. It also indicates that, in contrast to the source-edge temperature, the drain-edge temperature at the time of failure does depend greatly on the contact-to-gate spacings. Note also that the drain-edge temperature at the time of failure varies more than the source-edge temperature. These simulation results clearly indicate that it is not possible to define the onset of device failure, particularly the onset of second breakdown, in terms of a unique temperature value. Furthermore, accurate circuit simulation of ESD/EOS stress events in protection devices will require the simulation environment to account for steep temperature gradients. For example, the model for the parasitic bipolar npn transistor of an nMOS ESD protection device proposed in Section 6.1 uses the drain-edge temperature to evaluate avalanche breakdown model parameters and the source-edge temperature to determine the model parameters for the junction saturation currents and transistor gain.

6.3 SUMMARY

Experiments were performed on a typical I/O protection structure used in advanced CMOS processes to determine ESD-HBM and EOS failure thresholds in terms of power-to-failure and current-to-failure profiles. The performance was measured for various values of major layout parameters such as device width and contact-to-gate spacings. Device-level electrothermal simulations were done to explain the observed failure mechanisms. In spite of the fact that these simulations were 2D in nature, they provided qualitative explanations of the phenomena leading to device failures and correctly predicted the experimental trends of the EOS failure thresholds. For high-energy short-duration stress events, devices appear to fail as a result of the formation of poly filaments with no second breakdown effects. For other stress conditions, the devices display gross failure originating from device operation in the second breakdown regime. It appears that the second breakdown of the parasitic bipolar transistor occurs when the device has heated up to the point at which the effects on device conduction due to thermal carrier generation offset the effects of the mobility degradation and the reduction of the impact ionization rates. Thus, the time for the onset of second breakdown depends on the stress current level and the device's power dissipation and geometry. Simulation results also suggest the presence of very steep temperature gradients within the device. Thus, it is not possible to define the onset of second breakdown in terms of a unique temperature value. Simulation results presented here also suggest that the current vs. time-to-failure ($I_f - t_f$) relationship should be used as a measure of device hardness against EOS. When the EOS can be modeled as a constant current source, the device ability to handle the stress can be measured by the value of t_f.

The analysis of the effect of contact-to-gate spacing on the failure thresholds presented in this chapter together with uniform stress current distribution and heat removal considerations led to the development of a source contact placement scheme for improved ESD/EOS performance of typical I/O protection structures in MOS processes [21].

7

CIRCUIT-LEVEL ELECTROTHERMAL SIMULATION

The scope of previous work on electrothermal simulation using network analysis techniques has been limited due to the lack of avalanche breakdown models which can efficiently describe the temperature dynamics [59, 58, 89]. For EOS/ESD, an accurate description of temperature-dependent device electrical behaviour including breakdown phenomena is of central importance. This chapter presents the fundamental aspects of circuit-level electrothermal simulation and their implementation in the newly developed circuit simulator iETSIM. In Section 7.1, the temperature-dependent device parameters of the device models used in SPICE [6] are reviewed. This section also discusses a generalization of Miller's formula [72] to account for temperature-dependent avalanche breakdown phenomena. Section 7.2 presents an algorithm that enables circuit simulators to safely handle avalanche breakdown characteristics. Section 7.3 introduces a thermal model that enables efficient coupling of the thermal and electrical device phenomena for circuit simulation. With all of these elements in place, Section 7.4 introduces the electrothermal circuit simulator iETSIM.

7.1 TEMPERATURE EFFECTS AND DEVICE MODELS

Dominant temperature effects in device circuit models come from the temperature dependencies of silicon parameters such as the energy gap E_g, the effective density of states in the valence band N_v and conduction band N_c, the carrier mobility μ, the carrier diffusion coefficient D, and the impact ionization rates for electrons α_n and holes α_p. The energy gap and the effective density of states determine the temperature dependency of the intrinsic carrier concentration which appears in the expressions for the junction leakage currents. The following expressions are generally used to calculate those material parameters [90, 95, 18, 97]:

$$\begin{array}{lcll} E_g & = & 1.1255 - \frac{4.73\times10^{-4}T^2}{T+636} & eV \\ n_i & = & n_{io}\left(\frac{T}{T_o}\right)^{3/2} \exp\left(\frac{E_g}{2kT} - \frac{E_{go}}{2kT_o}\right) & 1/cm^3 \\ \mu & = & \mu_o\left(\frac{T}{T_o}\right)^{\nu} & cm^2/V-s \\ D & = & \mu\frac{kT}{q} = D_o\left(\frac{T}{T_o}\right)^{\xi} & cm^2/s \\ \alpha & = & \alpha_\infty \exp\left(-\frac{\mathcal{E}_{cr}}{\mathcal{E}}\right) & 1/cm \\ \mathcal{E}_{cr} & = & \frac{E_g}{q\lambda_o \tanh\left(Ep/2kT\right)} & V/cm \end{array}$$

where $\xi \equiv \nu + 1$, $\mathcal{E}$ is the electric field, Ep is the optical-phonon energy ($0.063\ eV$), and λ_o is the optical-phonon mean free path at room temperature.

7.1.1 Semiconductor Resistors

The temperature dependency of semiconductor resistors arises from the mobility and carrier concentration. Figure 7.1 shows the MEDICI-simulated [97] sheet resistance of the drain/source diffusions of the nMOS devices studied in Chapter 3. Initially, the resistance increases as a result of the mobility degradation due to increased acoustic-phonon (lattice) scattering. However, at high enough temperatures, the increase in thermal carrier generation becomes the dominant effect causing the resistance value to decrease. The value of the temperature T_{crit} at which the resistance begins to decrease is a function of the doping level; T_{crit} decreases as the doping level decreases.

Consider the semiconductor resistor shown in Fig. 7.2 under nonisothermal conditions. Assume that the time-dependent temperature profile is only a function of the spatial coordinate parallel to the direction of current flow and that the Joule heating of the resistor is negligible compared to the heat source determining the temperature profile (for example, a nearby reverse-biased junction carrying high currents). Under these conditions, the resistor will have a temperature-induced negative differential resistance behaviour when the following condition is satisfied:

$$\frac{dR}{dt} = \frac{1}{W}\int_0^L \frac{\partial R_{sh}}{\partial T}\frac{\partial T(x,t)}{\partial t}dx \leq 0.$$

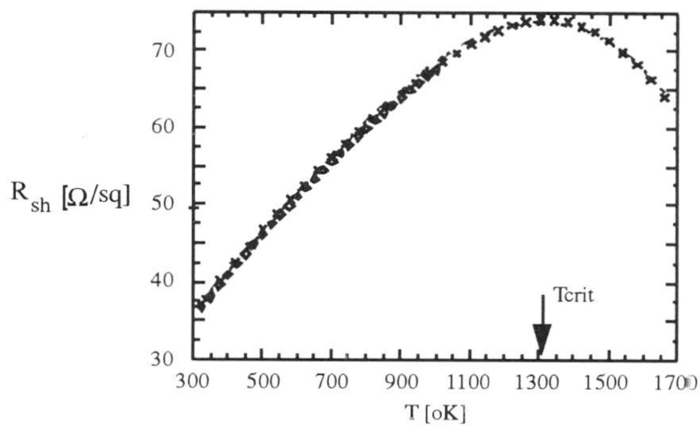

Figure 7.1 Drain/source sheet resistance for a 0.6 μm nMOS device.

Note that the time at which the condition is satisfied is a function of the resistor length, the sheet resistance and the temperature distribution. Furthermore, at the time the condition is satisfied, the portion of the resistor whose temperature is above T_{crit} decreases as the temperature gradient increases. (This is a much more general condition to define the onset of thermal instability in semiconductor resistors.) Under uniform temperature distribution, the condition given here reduces to the one given in [44]. Temperature-induced negative differential resistance can cause current localization leading to resistive thermal runaway under EOS/ESD conditions, particularly in a region with a low doping level.

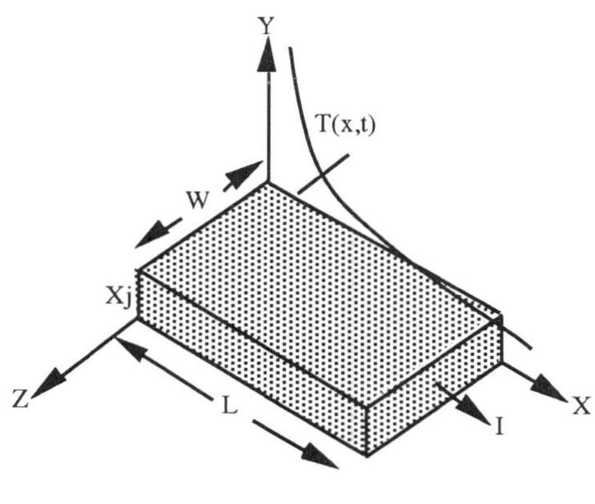

Figure 7.2 A semiconductor resistor under *1D* nonisothermal conditions.

Temperature-dependent semiconductor resistance can be described to a circuit simulator by a third-order polynomial

$$R_{sh}(T) = R_{sho} + (T - T_o)(m_1 + (T - T_o)(m_2 + m_3(T - T_o))).$$

For accurate circuit simulation of resistive breakdown, each resistor has to be split into a number of sections. Each section is assumed to be in thermal equilibrium at a temperature provided by the circuit simulation environment.

7.1.2 P-N junctions

The junction potential used in determining the junction capacitances is temperature dependent. It is given by [95]

$$\phi(T) = \frac{kT}{q} \ln\left(\frac{N_a N_d}{n_i(T)^2}\right) . \quad (7.1)$$

The temperature dependency of an ideal abrupt $p - n$ junction is given by [95]

$$\begin{aligned} i &= I_S(T)\left(\exp\left(qv/kT\right) - 1\right) \quad (7.2) \\ I_S(T) &= I_{SO}\left(\frac{T}{T_o}\right)^{(3+\xi/2)} \exp\left(\frac{E_g}{kT} - \frac{E_{go}}{kT_o}\right) \end{aligned}$$

where ξ was defined earlier as the exponent of the carrier-diffusion coefficient. This expression can be easily generalized to model nonabrupt junctions and to include high-level injection effects [95].

The avalanche breakdown phenomenon at a fixed temperature under low-level injection can be modeled using the empirical Miller multiplication factor M [72]

$$M = \frac{1}{1 - \left(\frac{v}{V_{BV}}\right)^n} \quad (7.3)$$

where V_{BV} is the junction breakdown voltage. Now, consider temperature-dependent avalanche breakdown phenomena. For a reverse-biased abrupt junction under low-level injection conditions, the impact ionization multiplication factor can be written as [78]

$$M - 1 = \int_0^W \alpha_\infty \exp\left(-\frac{\mathcal{E}_{cr}}{\mathcal{E}}\right) dx \quad (7.4)$$

where W is the width of the depletion region and $\mathcal{E}_{cr}$ is the critical field described earlier. Assuming that most of the carrier multiplication takes place near the junction, the integration can be carried over and results in [78, 54]

$$M - 1 = \frac{\alpha_\infty \mathcal{E}_m W}{\mathcal{E}_{cr}} \exp\left(-\frac{\mathcal{E}_{cr}}{\mathcal{E}_m}\right). \tag{7.5}$$

For an abrupt junction with built-in voltage V_{bi} [95]

$$\begin{aligned} \mathcal{E}_m &= \sqrt{2qN_b(v_{bi} - v)/\epsilon} \\ W &= \sqrt{2\epsilon(v_{bi} - v)/qN_b}\ . \end{aligned}$$

The temperature dependency of M comes from the value of $\mathcal{E}_{cr}$. Figure 7.3(a) shows a typical plot of 1-1/M versus the junction voltage at various temperatures. Shown in Fig. 7.3(b) are the temperature dependencies that should be implemented in Miller's parameters V_{BV} and n. For circuit simulation, these parameters can be fit with second-order polynomials.

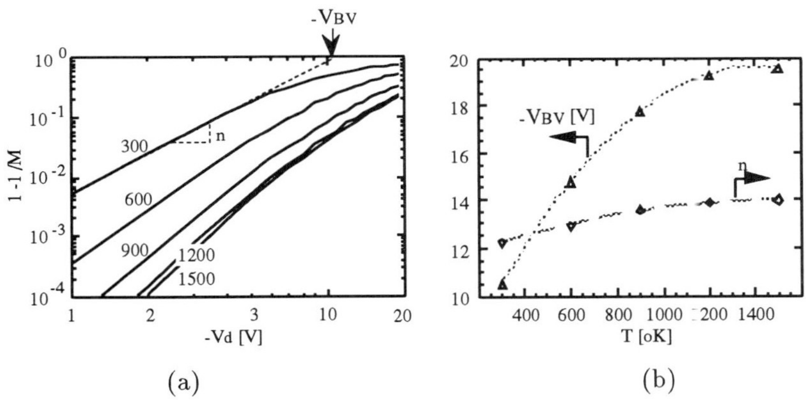

Figure 7.3 (a) Plot of $1 - 1/M$ vs. junction voltage under low-level injection in an abrupt junction with $N_b = 2 \times 10^{17}\ cm^{-3}$. The solid line shows how these curves define the parameters used in Miller's approximation of the multiplication factor. (b) Plot of the temperature-dependent breakdown voltage V_{BV} and the exponent n.

High-level injection phenomena in reverse-biased junctions are due to the space charge contributed by the mobile carriers. This is reflected by a reduction of the maximum electric field $\mathcal{E}_m$. This effect can be taken into account by replacing N_b by an effective doping [80, 78]

$$N_{beff} = N_b\left(1 + \frac{i}{q\nu_{sat} A N_b}\right) \tag{7.6}$$

where i is the junction current (negative for reverse bias), ν_{sat} is the saturated velocity of the mobile carriers, and A is the junction area. For circuit simulation, a first-order model for high-level injection effects in breakdown can be implemented by adding $r_{sc}i_d$ to the numerator of $\frac{v}{V_{BV}}$ in Miller's expression [80] with

$$r_{sc} \equiv \frac{W^2}{2\epsilon A \nu_{sat}}. \tag{7.7}$$

Results from the electrothermal diode model, including the avalanche breakdown and high-level injection effects, were compared with corresponding results from the device simulator TMA-PISCES [97]. The simulated diode structure, shown in Fig. 7.4(a), was created from the LDD nMOS devices studied in the previous chapter by removing one of the n diffusions. Figure 7.4(b) shows the predicted terminal voltage (reverse bias) versus diode temperature for different current levels under steady-state conditions according to the device simulator (symbols) and the lumped electrothermal model (solid lines). It can be seen that the lumped model closely follows the results of the *2D* device simulator.

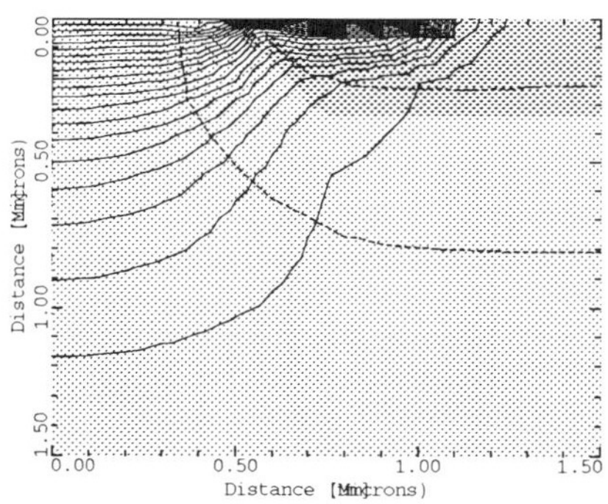

(a)

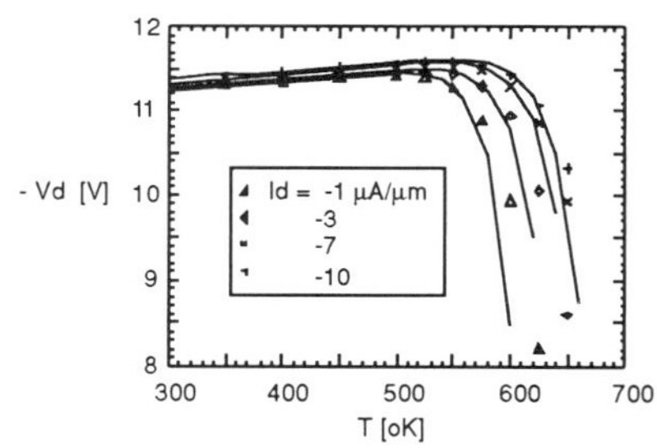

(b)

Figure 7.4 (a) Cross-section of the "LDD" n^+p diode showing the placement of the ohmic contacts and current flow lines; the flow lines spread at higher temperatures while the depletion region shrinks. (b) Steady-state terminal voltage versus junction temperature at different current levels. Symbols mark TMA-PISCES results. Solid line is the lumped electrothermal diode model result.

7.1.3 Bipolar and MOS Transistors

Dominant temperature-dependent effects for bipolar and MOS transistors can be accounted for by using the above results for the $p - n$ junction.

In bipolar transistors, the current gain is a function of temperature that can be modeled as [98]

$$\beta(T) = \beta_o \left(\frac{T}{T_o}\right)^{xtb} . \tag{7.8}$$

This expression is particularly appropriate for modeling the temperature dependencies of the current gain associated with the parasitic bipolar transistor of a typical ESD/EOS nMOS device, since in that case β is given by [93, 95]

$$\beta \approx 2kT\mu\tau/q{L_g}^2 , \tag{7.9}$$

where τ is the minority carrier lifetime in the substrate and L_g is the nMOS transistor channel length.

In MOS transistors, the temperature term appears explicitly in the surface mobility (transconductance) and in the junction and surface potentials.

It is important to keep in mind that the device model may be required to be dependent on more than one spatial temperature if the temperature gradients are steep. For example, the parasitic bipolar transistor of an nMOS device, subjected to EOS/ESD stress conditions, was shown to exhibit large temperature gradients. In this case, a good circuit level approximation of this phenomenon requires that model parameters related to impact ionization be controlled by the temperature of the drain junction near the gate edge, while other model parameters are controlled by the source temperature at the gate edge, for example.

7.2 SIMULATION OF AVALANCHE BREAKDOWN

Let us assume that we have a nonlinear resistor defined by $i = f(v)$. In the context of Modified Nodal Analysis (MNA), the stamps [1] for this nonlinear resistor can be written in terms of

$$g^{(k)} = \frac{df}{dv}\Big|_{v=v^{(k)}} \tag{7.10}$$

$$i^{(k)} = f(v^{(k)}) \tag{7.11}$$

where $v^{(k)}$ is the terminal voltage at the kth iteration of Newton's method [1]. A problem arises when the function f is not smooth and contains singularities as in the case of the multiplication factor used to model avalanche breakdown. In such cases, the simulation convergence may be destroyed. The numerical process for solving the nonlinear system could lead to a $v^{(k)}$ that does not belong to the domain of the function f thus preventing the iterative process from continuing. To overcome this problem, the following algorithm can be used to guarantee that the element's stamp represents a feasible operating condition regardless of the result obtained in the previous iteration. (The algorithm assumes that the element's branch current has been declared as a variable (i_v) in the MNA system of equations. The branch current is used to determine a feasible operating voltage $v_{tg}^{(k)}$ when $v^{(k)}$ is not.)

1. If $v^{(k)}$ is a physically admissible operating condition of the element, determine the stamp entries using
   $$g^{(k+1)} = \frac{df}{dv} \Big|_{v=v^{(k)}}$$
   $$i^{(k+1)} = f(v^{(k)}) \; .$$
2. Else,
   if $v_{tg}^{(k)} = f^{-1}(i_v^{(k)})$ exists, then determine the stamp entries using
   $$g^{(k+1)} = \frac{df}{dv} \Big|_{v=v_{tg}^{(k)}}$$
   $$i^{(k+1)} = i_v^{(k)} \; .$$
   else, set $v^{(k)} = 0$ and go to step 1.

Let us consider the application of this algorithm for the modeling of avalanche breakdown in diodes. In this case the diode characteristics can be expressed as [95]

$$i_d = I_s M(\exp(\frac{v_d}{v_{tj}}) - 1),$$

where v_{tj} is the thermal voltage of the junction and the multiplication factor is given by

$$M = \begin{cases} 1.0, & 0 < v_d \\ \frac{1}{1-(\frac{v_d}{BV})^n} & BV < v_d < 0. \end{cases}$$

The algorithm used to implement this diode model can be written as

```
vtj = kT/q;
Compute_Ideal_Diode_Stamp();
if (vdk < 0) {
```

```
        if (vdk > BV) {
            aux = exp ( n * log (vdk/BV) );
            M = 1.0 / (1.0-aux);
            ik = ik * M;
            gk = M * (gk + ik * n * aux / vdk);
        } else {
            if (id_v < 0.0) {
                ik = id_v;
                vdk = BV * exp ( log(1.0+IS/ik) /n);
                aux = exp (n * log(vdk/BV));
                M = 1.0 / (1.0-aux);
                gk = IS * exp (vdk/vtj) / vtj;
                gk = M * (gk + ik * n * aux / vdk);
            } else {
                /* This is really a last resort */
                M = 1.0;
                gk = IS/vtj;
                ik = 0.0;
            }
        }
    }
```

To include the avalanche breakdown effects of the bipolar transistor collector junction, the basic Gummel-Poon model used in SPICE [6] is modified as shown in Fig. 7.5(a) [78, 80]. The additional nonlinear current source i_{cm} can be expressed in terms of the device terminal voltages and collector current without avalanche (i_c) by $i_{cm} = (M-1)i_c$. The multiplication factor M is given by

$$M = \begin{cases} 1.0 & 0 < v_{bc} \\ \frac{1}{1-\left(\frac{v_{bc}}{BV_{cbo}}\right)^n} & BV_{cbo} < v_{bc} < 0 \end{cases}$$

where BV_{cbo} is the BC breakdown voltage with open emitter. The algorithm used to implement the bipolar transistor model with collector breakdown is written as

```
loop: Compute_Standard_Gummel_Poon_Model();
        if (vbc < 0.0) {
            if (vbc > BVcb) {
                aux = exp ( n * log(vbc/BVcb) );
```

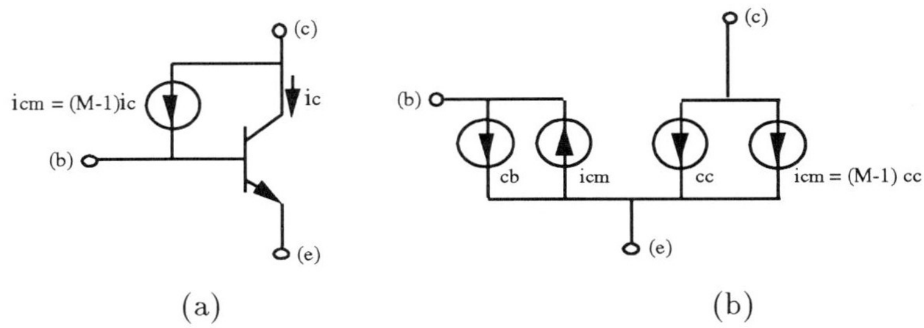

Figure 7.5 (a) An *npn* bipolar transistor with BC avalanche breakdown, and (b) the corresponding transport model. The current sources *cc* and *cb* are calculated according to the standard Gummel-Poon model (see Section 8.3.4).

```
            M = 1.0 / (1.0-aux);
            icm = (M-1) * cc;
            gcm = (M-1) * (go+gmu) - cc*n*aux*M*M/vbc;
        } else {
            if ((cc_v > cc) && (cc > 0.0)) {
                vbc = BVcb * exp ( log(1.0-cc/cc_v) / n );
                goto loop;
            } else {
                vbc = 0.9 * BVcb; /* 0.9 is a heuristic */
                goto loop;
            }
        }
    }
```

where *cc*, *go*, and *gmu* are the collector current and conductances (hybrid-π model) associated with the standard SPICE Gummel-Poon model [6].

With the avalanche breakdown model of the bipolar transistor defined, a breakdown model for an nMOS transistor can be implemented as shown in Fig. 7.6 [93, 52]. In this figure, the standard level-2 SPICE model for MOS transistors can be used to determine the nonbreakdown MOS currents. The substrate hole current induced by the channel hot electrons is modeled by [53]

$$i_{he} = \begin{cases} i_{ds}(v_{ds} - v_{sat}) \exp\left(-\frac{b_i l_c}{v_{ds} - v_{sat}}\right) & v_{ds} > v_{sat}, \\ 0 & otherwise \end{cases}$$

where v_{sat} is the drain saturation voltage and a_i, b_i, and l_c are ionization parameters described in [53].

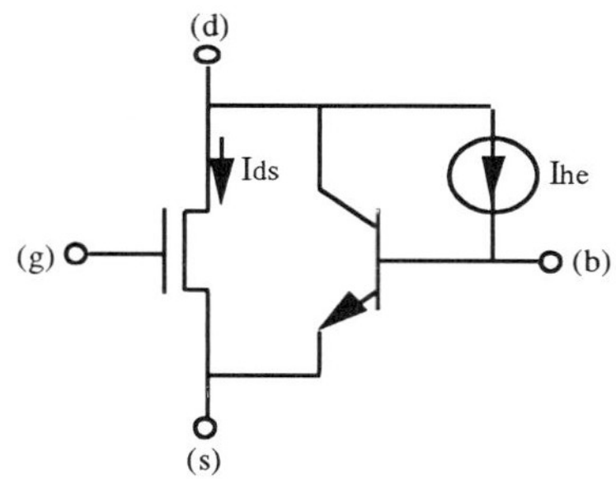

Figure 7.6 The *nMOS* transistor breakdown model with the substrate current and the parasitic bipolar transistor that incorporates the drain-substrate avalanche breakdown.

The algorithms presented in this section are simpler, more robust, and more efficient versions of the corresponding algorithms introduced in [26] that were developed using the iSMILE circuit simulation environment [101]. In particular, for the case of the bipolar transistor, the algorithm introduced here requires that only the collector current be declared a variable in contrast to the original version [26] where it was necessary to have the base current also declared as a variable in the MNA equations. The algorithms presented here were implemented in the model routines of the new electrothermal simulator iETSIM that will be introduced below.

7.3 TEMPERATURE MODEL FOR ELECTROTHERMAL SIMULATION

The scope of previous work on electrothermal simulation using network analysis techniques has been limited because of the methods and models used to describe the temperature dynamics [59, 58, 89]. In [59], a method to analyze a power bipolar transistor was introduced using the finite-difference approach to generate an electrical-analog of the *3D* heat diffusion equation which is then solved together with the electrical lumped device model using relaxation techniques. This method is not efficient for studies of the self-heating of circuits involving more than one device or when the temperature gradients are steep ($\sim 200\ ^oC/\mu m$) as in ESD stress of nMOS protection devices. In [89], the Green's function approach is used to obtain a steady-state closed-form expression of the temperature distribution that is solved iteratively together with an MOS breakdown model based on spline interpolating functions. This approach is limited to steady-state analysis because of the temperature model used. It should be also noted that the work in [45] was directed to implement low-frequency filters using the large time constants characteristic of the heat flow and that the electrothermal model used is not applicable to accurate EOS/ESD simulation. The thermal model used in [45], based on the Laplace transform approach, is limited to two-dimensional heat flow in the silicon chip and is not appropriate for efficient time-domain circuit simulation.

Here, a temperature model based on an electrical-analog implementation of the time-dependent *3D* heat diffusion equation solution given by [43] is used for circuit-level electrothermal simulation. Dwyer et al. [43] solved the *3D* heat diffusion equation for an infinite linear medium assuming that heat is generated uniformly within an $a \times b \times c$ parallelepiped. If the rate of heating per unit volume is $P(t)/\Delta$, where $P(t)$ is the instantaneous power dissipation of the device and $\Delta = abc$, the temperature distribution can be written as [43]

$$T(\vec{r},t) = T_o + \frac{1}{\rho c_p \Delta} \int_0^t P(\tau) G(x,a,\tau) G(y,b,\tau) G(z,c,\tau) d\tau \tag{7.12}$$

where

$$G(x,a,\tau) \equiv \frac{1}{2} \left\{ erf\left(\frac{a/2+x}{2\sqrt{\tau\kappa/\rho c_p}} \right) + erf\left(\frac{a/2-x}{2\sqrt{\tau\kappa/\rho c_p}} \right) \right\}$$

and $\vec{r}$ is the location of the measurement point with respect to the center of the heat source, κ is the thermal conductivity, c_p is the specific heat and ρ is the density. As an example, consider an nMOS protection device being stressed with its drain potential positive with respect to the source potential. Figure 7.7 shows the model of the heat source volume and a typical selection of the coordinate system which is assumed to coincide with the principal axis of the parallelepiped.

Equation (7.12) is valid for an infinite linear medium. For application to temperature modeling in actual ICs, certain considerations have to be given to the passivation layers and the temperature-dependent thermal parameters of the substrate. ICs have a passivation layer whose thermal conductivity is much lower than the substrate thermal conductivity [90, 22]. Therefore, the temperature increase given by the above expression would underestimate the actual value. This can be fixed by including an additional model parameter ζ so that the prefactor of the integral becomes $\zeta/\rho c_p \Delta$. This user-specified model parameter can take any value between 1 and 2. For instance, if the passivation

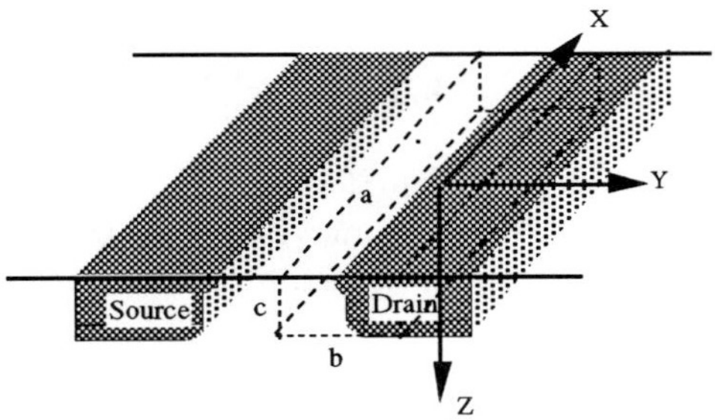

Figure 7.7 An nMOS protection device stressed drain positive with respect to source. The model of the heat source volume and a typical selection of the coordinate system which is assumed to coincide with the parallelepiped's principal axis.

layer is assumed to be a perfect thermal insulator, then $\zeta = 2$. Thermal parameters of the substrate material are temperature dependent. In silicon, the lattice thermal conductivity and the ρc_p product are given in terms of the local lattice temperature T by [90]

$$\begin{aligned} \kappa &= 1.5486 \left(\frac{T}{300}\right)^{-4/3} \quad [W/cm -^o K] \\ \rho c_p &= 1.574 \left(\frac{T}{300}\right)^{0.1} \quad [J/cm^3 -^o K]. \end{aligned} \tag{7.13}$$

Although the temperature dependency of the material's density-specific heat product ρc_p can be ignored for the temperatures of interest (the error is less than 12% at 1000 ^oK), the same cannot be said for the thermal conductivity. Using the nonlinear thermal simulator iTSIM [22, 30], one can show that the nonlinear transient thermal response under pulse power stressing can be approximated (over a broad range of times-to-failure) by a linear medium with a proper choice of the thermal parameters [25]. Ash [7] has also shown that the Wunsch-Bell relation ($P_f \propto 1/\sqrt{t_f}$) derived with constant thermal parameters still holds in nonlinear materials whose thermal parameters can be described by a special set of functions (these functions are appropriate only over a limited range of lattice temperatures for silicon, thus limiting the validity of Ash's proof). In silicon ICs under EOS/ESD stress events, a suitable choice of the thermal parameters to be used in Eq. (7.12) can be obtained from Eq. (7.13) by having $800 < T < 900\,^oK$. Some researches have used $T \approx 1100\,^oK$ in their studies of second breakdown [5].

7.4 IETSIM: AN ELECTROTHERMAL CIRCUIT-LEVEL SIMULATION TOOL

iETSIM is a circuit-level electrothermal simulation tool written in the *C* language running under the UNIX operating system. For detailed information on iETSIM and how to use it, the reader is referred to Chapter 8.

The program is based on the Modified Nodal Analysis approach for circuit simulation [1] and can perform both steady-state and transient analysis. For transient analysis, iETSIM uses either backward Euler or trapezoidal integration methods with fixed or variable time step control. The program offers the user the possibility of running both isothermal and nonisothermal simulations. For a nonisothermal simulation, device temperatures are determined from the device power dissipation using an equivalent circuit that models a closed-form solution of the 3D heat diffusion equation. This approach allows the circuit variables and device temperatures to be solved self-consistently under non-isothermal conditions.

In addition to the standard linear circuit elements such as resistors, capacitors, linear voltage and current sources, the simulator has built-in power monitors and power-to-differential-temperature converters that can be placed anywhere in the circuit to monitor device temperatures and feed them back to the device electrothermal models.

Currently, iETSIM provides dc electrothermal models for diodes, bipolar and MOS transistors that are capable of describing device behaviour up to the onset of second breakdown. iETSIM was initially targeted for the analysis of ESD/EOS in I/O protection circuits, but it can also be used to study other circuit performance variations resulting from device self-heating effects.

7.4.1 Temperature Modeling in iETSIM

iETSIM's capability to model self-heating effects on electrical device behaviour is based in part on an electrical-analog implementation of the solution of the *3D* heat diffusion equation. To achieve this, iETSIM provides the user with two special circuit elements, namely, a power monitor (P) and a time-dependent resistor (T) that convert power (numerically represented as a node voltage) to differential temperature (numerically represented as a branch current). This differential temperature can then be fed to an ideal integrator circuit [46] whose output voltage is numerically equivalent to a temperature. This electrical-analog implementation of Eq. (7.12) and its modifications discussed in Section 7.3 is solved simultaneously with the circuit under analysis thus providing full coupling of the electrical and thermal phenomena.

In iETSIM, the power monitor is implemented as a voltage-controlled current-source referenced to ground whose "transconductance" value is no longer fixed, but determined instead by the current flowing into the device whose heating is being monitored. The value of the device power dissipation is then numerically equal to the value of the voltage drop across the output nodes of the power monitor. Let us suppose that the device whose power is to be determined is connected between nodes i and j and that its current is a variable in the circuit's MNA equations with index m. Furthermore, assume that the output nodes of the power monitor are labeled k and l as shown in Fig. 7.8 and that the index n has been assigned to its output current. Then, the stamp for the MNA implementation of the power monitor for the $k+1$ iteration is given by

$$\begin{matrix} l \\ m \\ n \end{matrix} \begin{bmatrix} & & & & & 1 \\ & & & & & -1 \\ -i_m^{(k)} & i_m^{(k)} & 1 & -1 & -i_m^{(k)} & 0 \end{bmatrix} \begin{bmatrix} v_i \\ v_j \\ v_k \\ v_l \\ i_m \\ i_n \end{bmatrix} : \begin{bmatrix} \\ \\ -i_m^{(k)} v_{ij}^{(k)} \end{bmatrix} .$$

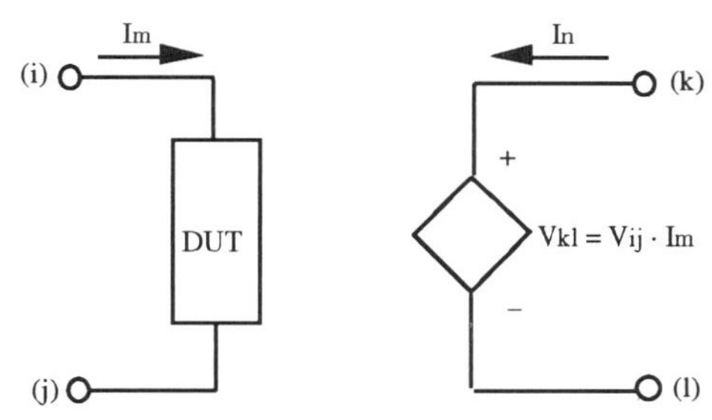

Figure 7.8 Schematic representation of the power monitor as implemented in iETSIM.

In iETSIM, the current through the special time-dependent resistor R is made to be numerically equivalent to the rate of change of the device temperature (at the location determined by the resistor parameters) when connected between the output of a device power monitor and ground (virtual ground when used as part of an integrator circuit [46]). The value of the resistor depends on the location (with respect to the center of the heat source) of the point whose temperature is being measured. The value of the special resistor is computed at each time point during the simulation according to

$$R(\vec{r},t) = \frac{8\rho C_p \Delta}{\zeta} \times \frac{1}{\left(erf(\frac{a/2+x}{2\sqrt{(D\tau)}}) + erf(\frac{a/2-x}{2\sqrt{(D\tau)}})\right)} \times \frac{1}{\left(erf(\frac{b/2+y}{2\sqrt{(D\tau)}}) + erf(\frac{b/2-y}{2\sqrt{(D\tau)}})\right)} \times \frac{1}{\left(erf(\frac{c/2+z}{2\sqrt{(D\tau)}}) + erf(\frac{c/2-z}{2\sqrt{(D\tau)}})\right)}. \quad (7.14)$$

In this expression, $D = \kappa/\rho c_p$ is the thermal diffusivity of the substrate. For computational speed, the error function, *erf(x)*, is approximated in iETSIM by a second-order polynomial (see Chapter 8).

As mentioned above, the power monitor and the time-dependent resistor are generally used in an integrator configuration as depicted in Fig. 7.9 to form an electrical analog of Eq. (7.12) that is solved jointly with the circuit under analysis.

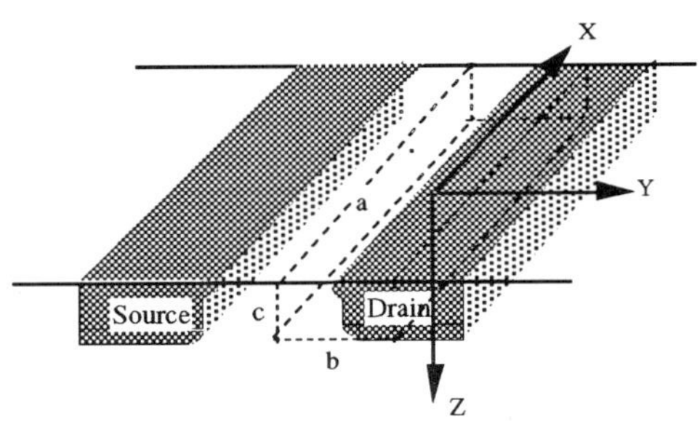

Figure 7.9 Arrangement of the power monitor and the time-dependent resistor in an integrator circuit to implement the electrical analog of the closed-form solution of the *3D* heat diffusion equation.

Note that the power monitor *should be connected* in such a way that any device power dissipation produces an increase of temperature (output voltage). This scheme is also appropriate to implement the temperature response at a point due to the superposition of different heat sources at different locations. For this, the user has to provide the corresponding power monitors and time-dependent resistors and to use the *current summing* property of the integrator input node.

7.4.2 Electrothermal Simulation of Temperature-dependent Nonlinear Resistors

Let a nonlinear resistor connected between circuit nodes i and j be represented by

$$i_{ij} = f(v_{ij}, T_1, T_2, \cdots)$$

where T_i represents temperature values. If the values of T_i are available as part of the MNA system, then the resistor stamp at the $k+1$ iteration can be written as

$$\begin{array}{c} i \\ j \end{array} \begin{bmatrix} \frac{\partial f}{\partial v_{ij}^{(k)}} & -\frac{\partial f}{\partial v_{ij}^{(k)}} & \frac{\partial f}{\partial T_1^{(k)}} & \frac{\partial f}{\partial T_2^{(k)}} & \cdots \\ -\frac{\partial f}{\partial v_{ij}^{(k)}} & \frac{\partial f}{\partial v_{ij}^{(k)}} & -\frac{\partial f}{\partial T_1^{(k)}} & -\frac{\partial f}{\partial T_2^{(k)}} & \cdots \end{bmatrix} \begin{bmatrix} v_i \\ v_j \\ T_1 \\ T_2 \\ \vdots \end{bmatrix} \tag{7.15}$$

$$= \begin{bmatrix} -i_{ij}^{(k)} + \frac{\partial f}{\partial v_{ij}^{(k)}} v_{ij}^{(k)} + \frac{\partial f}{\partial T_1^{(k)}} T_1^{(k)} + \frac{\partial f}{\partial T^{(k)}} T_1^{(k)} + \cdots \\ i_{ij}^{(k)} - \frac{\partial f}{\partial v_{ij}^{(k)}} v_{ij}^{(k)} - \frac{\partial f}{\partial T_1^{(k)}} T_1^{(k)} - \frac{\partial f}{\partial T_1^{(k)}} T_1^{(k)} - \cdots \end{bmatrix}.$$

In a purely electrical simulation environment such as in SPICE, the stamp for this nonlinear resistor would have entries only in columns 1 and 2. The additional stamp entries in the columns $\{T_1, T_2, \cdots\}$ and their corresponding terms in the *right-hand side* are required by the electrothermal simulation environment to complete the coupling of the temperature and the electrical models. These entries assure the convergence of Newton's method even under highly non-isothermal conditions.

In the case of the n^+p diode model including avalanche breakdown, one can express the diode current as

$$i_d = I_s(T_1)M(v_d, T_2)\left(\exp\left(qv_d/kT_2\right) - 1\right) \tag{7.16}$$

where T_1 is the temperature at the edge of the depletion region on the p side, T_2 is the junction temperature, and M is the impact ionization multiplication factor given above. The additional stamp entries required to couple the temperature model to the diode electrical model can be written as

$$\frac{\partial f}{\partial T_1^{(k)}} = i_d^{(k)} \left[\frac{1}{I_s} \frac{\partial I_s}{\partial T_1^{(k)}} - \frac{qv_d}{kT_1} \frac{\exp\left(qv_d/kT_2\right)}{T_1\left(\exp\left(qv_d/kT_2\right) - 1\right)} \right]$$

$$\frac{\partial f}{\partial T_2^{(k)}} = \frac{1}{M}\frac{\partial M}{\partial T_2} .$$

For the case of a bipolar transistor, such as the parasitic lateral bipolar in an nMOS ESD protection device, T_1 may correspond to the neutral-base temperature (near the gate edge of the source) controlling the thermal carrier generation and gain of the bipolar transistor, while T_2 may represent the temperature at the collector junction (gate edge of the drain) controlling the impact ionization rates. The full electrothermal stamp for the bipolar transistor with collector avalanche breakdown is given by

$$\begin{aligned} cb &= cb^{(k)} + g_\pi(v_{BE} - v_{BE}^{(k)}) + g_\mu(v_{BC} - v_{BC}^{(k)}) + \frac{\partial cb}{\partial T_1}(T_1 - T_1^{(k)}) \\ cc &= cc^{(k)} + (g_m + g_\mu)((v_{BE} - v_{BE}^{(k)}) - (g_o + g_\mu)(v_{BC} - v_{BC}^{(k)}) \\ &\quad + \frac{\partial cc}{\partial T_1}(T_1 - T_1^{(k)}) \\ icm &= icm^{(k)} + \left(cc^{(k)}\frac{\partial(M-1)}{\partial v_{CB}} + (M-1)\frac{\partial cc}{\partial v_{CB}}\right)(v_{CB} - v_{CB}^{(k)}) \\ &\quad + (M-1)\frac{\partial cc}{\partial T_1}(T_1 - T_1^{(k)}) + cc^{(k)}\frac{\partial(M-1)}{\partial T_2}(T_2 - T_2^{(k)}) \end{aligned}$$

where cc, cb, g_π, g_m, g_o, and g_μ are the collector and base currents and the conductances of the hybrid-π model associated with the standard SPICE Gummel-Poon model [6]. The collector avalanche current and multiplication factor are icm and M, respectively (see Fig. 7.5(b)).

Temperature-dependent device models implemented together with a temperature model, as described in the previous section, provide a natural fully coupled electrothermal description of the device behaviour that enables device simulation even under stringent nonisothermal conditions as is the case of devices subjected to transient EOS/ESD events.

7.4.3 ESD/EOS simulation examples

Figure 7.10 shows the simulation results for a 120 μm wide thick field oxide device driven into second breakdown by a 1.5 A current pulse. The device channel length is 1.6 μm. In this example, the temperature at the gate edge of the drain controlled the multiplication factor (impact ionization); the temperature at 0.2 μm from the midchannel point towards the source region determined the leakage currents and the gain of the parasitic bipolar transistor.

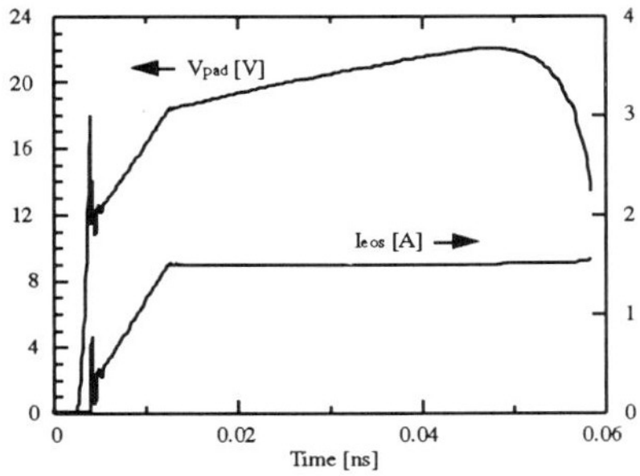

(a)

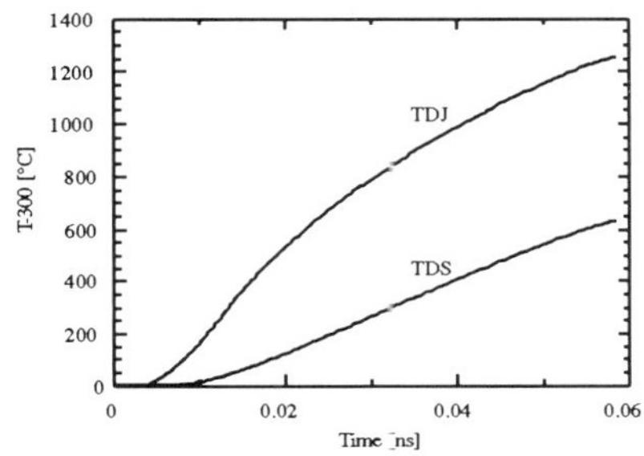

(b)

Figure 7.10 Simulated response of a $120\mu m/1.6\mu m$ thick field-oxide nMOS transistor to a 1.8 A EOS current pulse. (a) Drain (pad) voltage and drain (stress) current. The device is shown to go into second breakdown at about 50 ns. (b) Temperature at the gate edge of the drain junction (TDJ), and temperature at 0.2 μm from the midchannel towards the source region (TDS).

Initially, the device moves quickly from the first breakdown condition into the snapback region. As the stress current is ramped up, the pad voltage increases mainly as a result of the increased voltage drop across the drain and source contact resistances. After 1 ns, the increase in the pad voltage is a result of temperature increase that causes the current gain to decrease and the $BVcbo$ to increase. The current gain[1] decreases due to the mobility degradation. The value of $BVcbo$ increases due to the reduction of the impact ionization rates [18]. Near 50 ns, the device has heated up such that the effect of the thermal carrier generation on the device conduction offsets the degradation of the current gain and the reduction of ionization rates. This marks the onset of second breakdown for the device. It should be mentioned that this result matches well with the corresponding experimental data [17].

Figure 7.11 shows the predicted response of the above device to an 800 V HBM-ESD event. The circuit model parameters of the HBM-ESD were taken from to Table 1.4. Figure 7.11(a) shows the predicted pad voltage and ESD stress current waveforms. Figure 7.11(b) is a plot of the device temperatures: TDS is at the gate edge of the drain junction and TDS is the temperature at 0.2 μm from the midchannel towards the source region. The iETSIM input file used in this example is shown in Fig. 7.12.

[1] The current gain of the parasitic bipolar transistor is proportional to the mobility of the minority carriers in the channel region [93, 95].

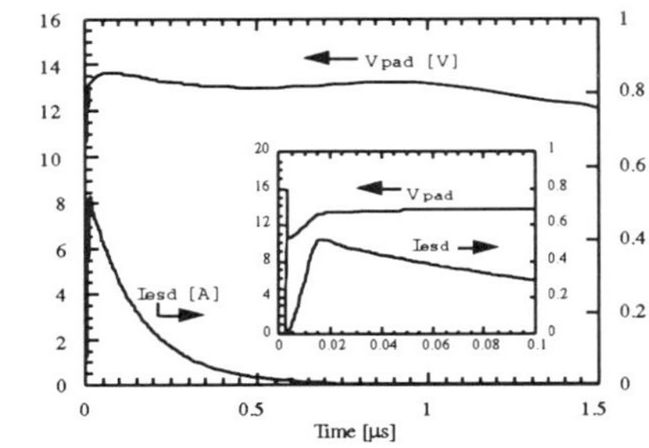

(a)

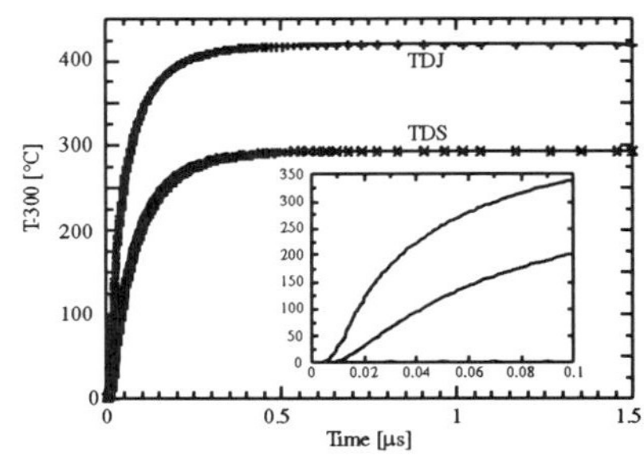

(b)

Figure 7.11 Simulated response of a $120\mu m/0.8\mu m$ thick-field oxide nMOS transistor to an 800 V HBM-ESD event. (a) Drain (pad) voltage and drain (stress) current. (b) Temperature at the gate edge of the drain junction (TDJ), and temperature at 0.2 μm from the midchannel towards the source region (TDS).

Finally, consider the gate-coupled nMOS devices described in Sections 2.5 and 3.3. For simulation purposes, it is assumed that the stress current is uniformly distributed among the different fingers in the thin oxide nMOS device. (This is a valid assumption for these devices according to results of [35].) Figure 7.13(a) is a SEM photograph (after stress and deprocessing down to polysilicon) of an 8-finger, 500 μm wide nMOS device with $CGS = 6$ μm. Figure 7.13(b) is a schematic cross-section of half of this 8-finger nMOS device. Figure 7.13(b) depicts the heat sources for the temperature model, the parasitic bipolar devices associated with each finger, and a blow-up of one of the two central fingers indicating the locations at which the device temperatures are monitored. The temperatures at these points are determined by the self-heating of the finger plus the contributions from the heat generated in the other fingers. For the temperature model, the power dissipation in each finger is assumed to take place within a parallelepiped whose dimensions are determined by the layout and the process. In Fig. 7.13(b), a is related to the width of the finger (a layout parameter) and c is equal to the junction depth (a process parameter, in this case $0.35\,\mu m$); b is a model parameter that was set to 0.4 μm based on temperature profiles obtained from 2D electrothermal simulations for different stress levels at the time of second breakdown as shown in Fig 6.6. For the simulation results described here on, the silicon thermal parameters were assumed constant calculated at 800 ^oK and the empirical parameter ζ in the temperature model was set to 1.75. The electrical parameters of the parasitic bipolar transistor were determined from experimental $I-V$ characteristics obtained using transmission line pulsing techniques. The iETSIM input file used to simulate the structure shown in Fig. 7.13 is shown in Fig. 7.14.

Figure 7.15 shows iETSIM simulation results for a stress current pulse of 3.0 A applied to the device in Fig. 7.13. Figure 7.15(a) indicates that the device reaches the onset of second breakdown (i.e. the onset of failure) at about 1.4 μs. As the stress current is ramped up (in 1 ns), the device moves quickly from the first breakdown condition into the snapback region (at nearly the $t = 0$ point in the graph). The pad voltage increases due to the increased voltage drop across the drain and source contact resistances. After 1 ns, the increase in the pad voltage is a result of temperature increase that causes the current gain to decrease and the $BVcbo$ to increase. The current gain decreases due to the mobility degradation. The $BVcbo$ increases due to the reduction of the impact ionization rates [18]. Near 1.4 μs, the device has heated up such that the effect of the thermal carrier generation on the device conduction offsets the degradation of the current gain and the reduction of ionization rates. This marks the onset of second breakdown for the device that is associated with the device failure. Figure 7.15(b) shows the simulated temperature rise at the two monitoring points shown in Fig. 7.13, the trace labeled TFC in this figure is the temperature rise contributed by the other device fingers at these two locations; clearly, the thermal coupling becomes important at about 1 μs.

Several iETSIM simulations of this kind were run for different levels of the stress current and layout parameters such as device width and contact-to-gate spacings. The corresponding simulation results were plotted in terms of the current profiles (level of stress current vs. time-to-failure) and compared with the experimental results reported in Section 6.1. These plots are shown in Figs. 7.16(a) and 7.16(b) where the solid lines represent iETSIM simulation results. As can be seen from these figures, the simulated current-to-failure profiles closely match the experimental data over a broad range of times-to-failure for the various combinations of layout parameters. Note that in contrast to the $2D$ iTSIM and TMA-MEDICI results presented in Chapters 3 and 6, iETSIM simulations do not understimate the device failure characteristics for short duration events. This can be attributed to the $3D$ built-in temperature model used in the electrothermal circuit simulator iETSIM.

7.5 SUMMARY

This chapter introduced device models suitable for electrothermal circuit simulation up to the onset of circuit thermal instability. Thermal instability of a stressed circuit or device is shown to be the result of either thermally induced negative differential resistance (NDR) in resistive regions, or due to junction second breakdown. It is shown that under typical ESD/EOS stress events (transient in nature), it is not possible to define a critical temperature at which thermal instability takes place. Thermally induced NDR takes place in semiconductor resistors when the thermal carrier generation offsets the mobility reduction; it can be a dominant effect in semiconductor resistances made in low doping regions and exposed to nonisothermal transient conditions. A general criterion for thermal instability of semiconductor resistors was introduced in terms of the sheet resistance and the temperature profile. In reverse-biased semiconductor junctions under transient nonisothermal conditions, second breakdown occurs at the time when thermal carrier generation becomes high enough to offset the effects of the mobility degradation and the reduction of the impact ionization rates. Under these circumstances, the time for the onset of second breakdown depends on the stress level and the device power dissipation per unit volume.

SPICE models for the bipolar (dc Gummel-Poon) and MOS (dc level 2) transistors were enhanced to account for temperature-dependent avalanche breakdown phenomena by using a generalization of Miller's expression for the multiplication factor.

The electrothermal models introduced here were implemented in the new electrothermal circuit simulator, iETSIM, that uses a closed-form solution of the heat diffusion equation to model the temperature behaviour in terms of device power dissipation. The device model routines implemented in iETSIM make use of an algorithm developed to guarantee the convergence of Newton's method for circuits having device models that include avalanche breakdown description. iETSIM's ability to capture the dominant physical phenomena leading to device failure (thermal breakdown) in advanced nMOS I/O devices under EOS events was demonstrated over a broad range of times-to-failure.

```
* This is a LSI's TFOD3 parasitic npn. W = 120 um, Xj = 0.3, L = 1.6 um.
*
* The HBM-ESD circuit model. The stress level is 800 V.
*
vc 25 0 step 0 800 2.5e-9 1.0e-8
cc 25 26 100e-12
ls 26 27 5e-6
rs 27 1 1500
cs 27 1 1e-12
vct 1 2 dc 0.0
*
* The protection device
*
.model qn npn BF=7.72 BVcb=-17.5 n0=3.5 is=1e-15 br=7.72 isc=1e-15 ise=1e-15
RC 2 3 2.5
Q1 3 4 5 8 10 qn vct
RB 4 0 5000
RE 5 0 2.5
*
* The temperature monitor
*
P1 0 6 3 0 vct
T1 6 7 3.6e-5 6.12e-3 6.12e-3 5.10e-5 5.10e-5 1.53e-5 1.53e-5  /* 120x0.3x1.0*/
C1 7 8 1e-6
E0 8 0 7 0 -10000
T2 6 9 3.6e-5 6.12e-3 6.12e-3 -1.02e-4 2.55e-4 1.53e-5 1.53e-5 /* at -1.0 */
C2 9 10 1e-6
R9 9 0 1e15
E0 10 0 9 0 -10000
*
.tran 0.0 1.5e-6 1e-9
.print tran v(2) i(vct)
.print tran v(8) v(10)
*
.xgraph
*
.options itlim=50 maxtsf=100 dx=1e-9 maxpts=1500
*
.end
```

Figure 7.12 iETSIM input file used to simulate the 800 V HBM-ESD event on a $120\mu m/1.6\mu m$ thick field-oxide nMOS transistor.

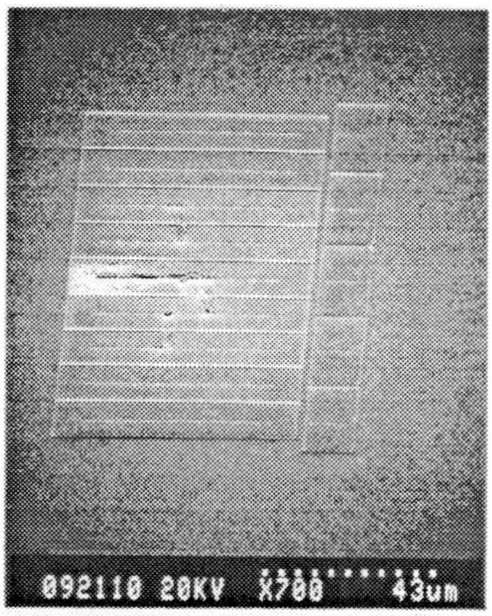

(a)

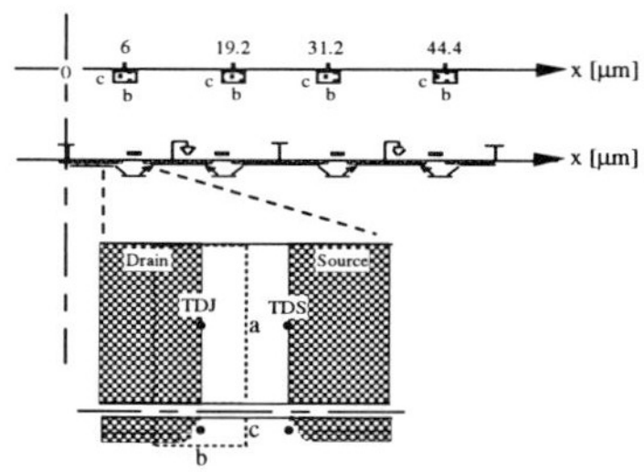

(b)

Figure 7.13 (a) SEM photograph taken, after deprocessing down to polysilicon, of an 8-finger gate-coupled nMOS transistor with 6 μm contact-to-gate spacing (CGS). Each finger is 62.5 μm wide with 0.6 μm channel length. The FOD was $50\mu m/1.2\mu m$. The device was stressed with a 100 μs, 0.82 A current pulse. (b) Schematic cross-section of half of this nMOS transistor that shows the relative locations of the heat sources for the temperature model, the drain/source contacts, the parasitic bipolar transistors, and the points (TDJ, TDS) whose temperature is consistently solved with the transistor electrothermal model.

```
* THIS DECK DETERMINES CURRENT PROFILES OF A 500 UM WIDE GATE-COUPLED DEVICE.
* THE DEVICE IS LAYOUT IN WITH 8 FINGERS 62.5/0.6 EACH. CGS = 6 UM
*
.MODEL TMOD HSMP A=62.5 B=0.4 C=0.35 TLIN=800 ZETA=1.75
*
.MODEL QN NPN BF=50.0 BVCB=-14 N0=4 IS=7E-16 ISC=1E-15 ISE=1E-15
+ BR=50.0 XTB=-1.5
*
IC 0 1 STEP 0 1.5 2.5E-9 1.0E-8
VCT 1 2 DC 0.0
*
RC 2 3 0.6
Q1 3 4 5 88 110 QN VCT
RB 4 0 5000
RE 5 0 0.6
*
P1 0 66 3 4 VCT    /* MONITORING POWER DISSIPATION OF Q1 ONLY */
E1 6 0 66 0 0.125  /* ASSUMED UNIFORM POWER DENSITY FOR ALL FINGERS */
*
T1 6 7 TMOD 1E6 0 0 6.0 6.0 0 0 /* DRAIN JUNC */
C1 7 8 1E-6
RIA1 7 0 1E15
EOA1 8 0 7 0 -10000
T2 6 9 TMOD 1E6 0 0 6.0 6.6 0 0 /* SOURCE JUNC */
C2 9 10 1E-6
RIA2 9 0 1E15
EOA2 10 0 9 0 -10000
*
* THE THERMAL COUPLING WITH OTHER FINGERS IS DETERMINED HERE
*
T3 6 11 TMOD 1E6 0 0 -6.0 6.0 0 0
T4 6 11 TMOD 1E6 0 0 19.2 6.0 0 0
T5 6 11 TMOD 5E5 0 0 31.2 6.0 0 0
T6 6 11 TMOD 1E6 0 0 44.4 6.0 0 0
T7 6 11 TMOD 1E6 0 0 -31.2 6.0 0 0
T8 6 11 TMOD 5E5 0 0 56.4 6.0 0 0
C3 11 12 1E-6
RIA3 11 0 1E15
EOA3 12 0 11 0 -10000
*
ETDJ 88 12 8 0 1.0
ETDS 110 12 10 0 1.0
*
.TRAN 0.0 1.0E-3 0.2E-9
.PRINT TRAN V(2) I(VCT)
.PRINT TRAN V(8) V(10) V(12)
.XGRAPH
*
.OPTIONS ITLIM=100 MAXPTS=1000 MINTSF=0.0005
.end
```

Figure 7.14 iETSIM input file used to simulate the gate-coupled nMOS device having 8 fingers each 62.5 μm wide with channel length of 0.6 μm and CGS of 6 μm.

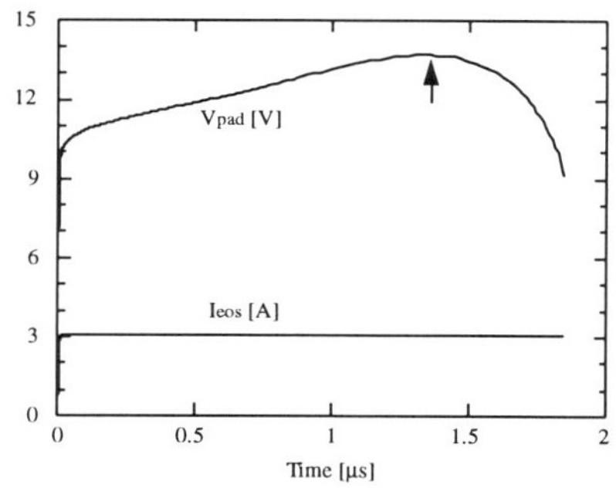

(a)

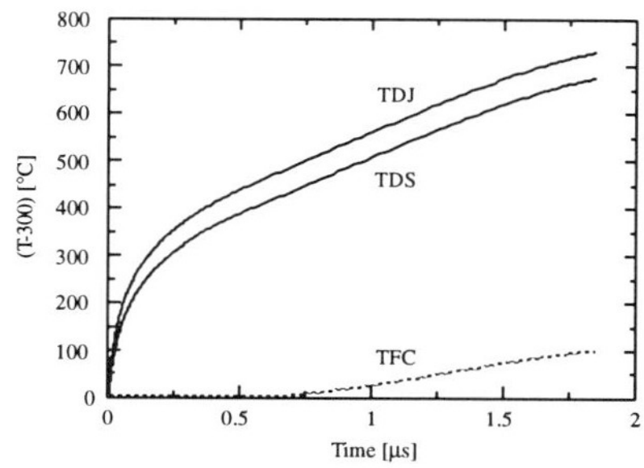

(b)

Figure 7.15 iETSIM's transient simulation results for the gate-coupled device in Fig. 7.13 subjected to 3.0 A of stress current. (a) The terminal current and voltage waveforms. (b) Transient device temperatures at the monitored points. (The TFC trace represents the temperature rise at these locations due to thermal coupling from fingers other than the central ones.)

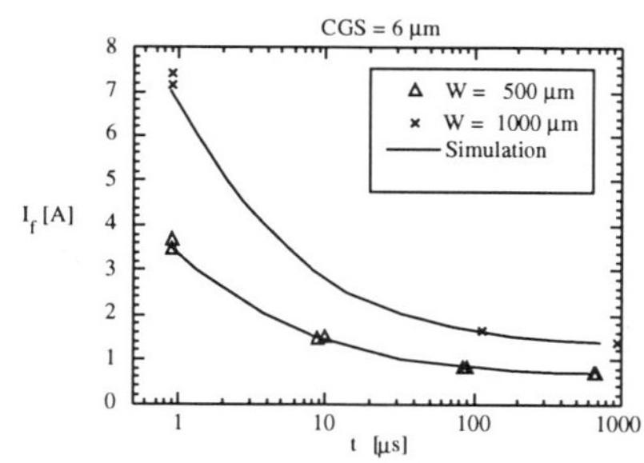

(a)

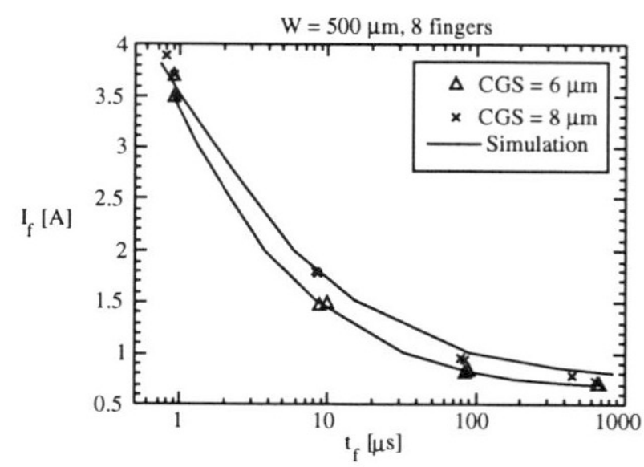

(b)

Figure 7.16 (a) Experimental and simulated EOS failure current profiles for gate-coupled devices with $CGS = 6\ \mu m$ but different effective widths. (b) Experimental and simulated EOS failure current profiles for a 500 μm wide gate-coupled device and for two different values of the CGS.

8

IETSIM : AN ELECTROTHERMAL CIRCUIT SIMULATOR

8.1 INTRODUCTION

iETSIM[1] is a circuit-level electrothermal simulation tool written in the *C* language running under the UNIX operating system. iETSIM is a self-contained system provided in 26 source files with a *makefile* to compile it and install the executable binary file *etsim*. The program is based on the Modified Nodal Analysis approach for circuit simulation and can perform both steady-state and transient analysis. Temporal discretization in iETSIM is done using either the backward Euler or the trapezoidal rule schemes. The resulting nonlinear discrete system of equations is solved using Newton's method. The sparse matrix solver in [57] is used to solve the linearized system.

The program offers the user the possibility of running both isothermal and nonisothermal simulations; device temperatures are determined from the device power dissipation using an equivalent circuit that models a closed-form solution of the *3D* heat diffusion equation. This approach allows the circuit variables and device temperatures to be solved self-consistently under nonisothermal conditions.

In addition to the standard linear circuit elements such as resistors, capacitors, linear voltage and current sources, the simulator has built-in power monitors and power-to-differential-temperature converters that can be placed anywhere in the circuit to monitor device temperatures and feed them back to the device's electrothermal models.

iETSIM provides dc electrothermal models for diodes, bipolar and MOS tran-

[1] The source code can be obtained upon request from the University of Illinois.

sistors that are capable of describing device behaviour up to the onset of second breakdown. iETSIM was initially targeted for the analysis of ESD/EOS in I/O protection circuits, but it can also be used for studying circuit performance variations resulting from device self-heating effects.

The core of the simulator is device independent, and as such it can be used to implement any kind of user-defined model. The user has to know the basics of C programming and Modified Nodal Analysis techniques to accomplish this task. The user first has to generate two data structures, namely, one for the model parameters and the other to hold device nodes and matrix pointers (devckt.h, diode.h, bjt.h). The user has to create a *C* source file that contains a routine to read a typical device input card, a routine to read the device model card, a routine that computes the linearized model and stamp entries, and a setup routine that sets all of the matrix pointers (see for example: cond.c, diode.c, bjt.c). Once these two files are ready, only a minor number of additions have to be made to the following files:

- devckt.h to make the new device known to the circuit data structure (Ckt_sta) and the model parameter data structure (Dev_Mod),
- etsim.c to allocate memory for the device data structure (near line 77),
- matrix.c in the routine *Set_Up_Ckt* for enforcing pointer initialization, and
- *lin* for actually calling the enforcing calls to the device linearization routine.

If the device is time dependent, proper calls should be added in update.c. Although this is far from being a complete guide to implementing a new device model in iETSIM, it is good overview for a knowledgeable CAD user who attempts to accomplish such a task.

8.2 RUNNING THE PROGRAM

To run iETSIM the user requires a circuit input file (SPICE-like) that describes the circuit and the tasks to be performed by the simulator. The syntax of that input file is described in the next sections. The generic command line syntax of iETSIM is

etsim [-b] [-v] [-t] inputfile [outfile]

The output file name is optional; if not given, the output will be directed to the standard output. If no flags are provided in the command line, the simulator will use the trapezoidal integration method for transient analysis with variable time step. The flag *[-b]* would instruct the simulator to use the backward Euler integration method with user-provided fixed time step for transient analysis. The flag *[-v]* would instruct the simulator to use the backward Euler integration method with variable time step for transient analysis. The flag *[-t]* would force the simulator to use the trapezoidal integration method with fixed time step for transient analysis.

8.3 INPUT FILE: CIRCUIT DESCRIPTION AND FORMAT

The syntax of the input file is very similar to a typical SPICE input deck. The input file for iETSIM describes the circuit topology and contains the circuit element values and parameters. Additionally, the input file could contain lines specifying the type of analysis, output format, and simulation control options. As in SPICE, the first line is always treated as a title, any line that begins with an $*$ is treated as a comment line, and all lines following the optional *.END* card are ignored. All other lines can be placed in arbitrary order. iETSIM's input parser is not case sensitive and expects tokens of information in a line to be separated by a pair of characters from the set of blank spaces, equal sign, and left and right parentheses. iETSIM recognizes only continuation lines in model cards. In this case, a '+' sign should appear in the first column of the continuation line. Name fields in the input deck can be alphanumeric, should not be longer than 10 characters, and should not contain any delimiters. Numeric fields can be either integer fields or floating-point fields possibly followed by an integer exponent. No alphanumeric scale factors in the numeric fields, such as *Meg, m, u, n, and p*, are recognized in the current implementation of iETSIM.

All circuit element lines begin with the element name, whose first letter defines the element type, followed by the element's nodes and parameters as described below. The ground node should be numbered as *'0'*. iETSIM uses the associated reference convention, i.e., current flows in the direction of the voltage drop. As in any MNA circuit simulator, the circuit cannot contain loops of voltage sources and/or inductors and cannot contain cutsets of current sources and/or capacitors. Each node should have a dc path to ground and should be connected to at least one circuit element.

8.3.1 Element Lines

In iETSIM, the user has access to standard linear passive circuit elements and linear sources. The line syntax for each of these elements is described in the next sections.

Passive elements

Currently, iETSIM recognizes linear time-independent resistors, capacitors and inductors with the following line syntax:

RXXXXX N+ N- VALUE
CXXXXX N+ N- VALUE
LXXXXX N+ N- VALUE.

Examples:

r1 1 2 1000
cbc 4 5 1.0e-9
Ld1 4 0 1e-6

Independent current and voltage sources

The general form of the line syntax describing these sources is

IXX N+ N- TYPE VALUE IH TD TR TAU
VXX N+ N- TYPE VALUE VH TD TR TAU

iETSIM supports three different types of independent sources, namely, TYPE can be DC, EXP, or STEP. For dc sources, VALUE must be specified. For exponential sources, VALUE represents the low level initial value (LO) of the source, HI is the peak value, TD is the delay time, TR is the rise time, and TAU is the decay time. For step sources, specification of the TAU parameter is not required. For the time-dependent sources, the value is computed according to

$$f(t) = \begin{cases} LO & \text{if } t \leq TD \\ LO + \frac{HI-LO}{TR}(t - TD) & \text{if } TD < t \leq TD + TR \\ HI & \text{if } TD + TR < t, \text{ step source} \\ (HI - LO) * e^{-t-TD-TR/TAU} + LOW & \text{if } TD + TR < t, \text{ exp. source} \end{cases}$$

Examples:

vcc 1 0 dc 5
I1 2 4 step 0.0 1.0 1e-9 1.0e-8
v2 6 3 exp 0.0 1.0 1e-9 1.0e-8 1.5e-6

8.3.2 Linear Dependent Sources

In iETSIM the user has access to various linear time-independent controlled sources such as current-controlled current sources (F), current-controlled voltage sources (H), voltage-controlled current sources (G), and voltage-controlled voltage sources (E). The generic line syntax for these linear sources is

FXXXXX N+ N- NAME VALUE
HXXXXX N+ N- NAME VALUE
GXXXXX N+ N- NC+ NC- VALUE
EXXXXX N+ N- NC+ NC- VALUE

where N+ and N- are the positive and negative nodes, respectively, NAME is the name of a circuit voltage source or inductor whose current is the controlling current (necessarily a variable in the MNA context), and NC+ and NC- are the nodes whose voltage difference is the control variable.

Examples:

F1 10 4 L1 0.5
Hx 5 7 vesd 2.0
G1 0 2 4 12 1.0e-4
E1 12 9 7 8 1e4

8.3.3 Special Elements for Temperature Monitoring

The iETSIM capability to model self-heating effects on device electrical behaviour is based on an electrical-analog implementation of the solution of the *3D* heat diffusion equation. To achieve this, iETSIM provides the user with two special circuit elements, namely, a power monitor (P) and a time-dependent resistor (T) that convert power (numerically represented as a node voltage) to differential temperature (numerically represented as a branch current). The generic line syntax of these elements is

PXXXXX N+ N- NC+ NC- IDEV
TXXXXX N+ N- MNAME RO XS XM YS YM ZS ZM

Note that the power monitor is a generalization of a voltage-controlled current source (G) whose "transconductance" value is no longer fixed but determined by the current flowing into the device whose heating is being monitored. The value of the device power dissipation is numerically equal to the value of the voltage drop across the N+ and N- nodes of the power monitor. NC+ and NC- are the device nodes whose voltage difference is used to determine the device power dissipation. IDEV is the name of an inductor or voltage source connected in series with the device whose power dissipation is being determined.

The current through the special time-dependent resistor (T) is numerically equivalent to the rate of change of the device temperature (at the location determined by the resistor parameters) when connected between the output of the device power monitor and ground (virtual ground when used as part of an integrator circuit [46]). The value of the resistor is computed at each time point according to (see Chapter 7)

$$R(t) = \frac{R}{G(xms, a, t)G(yms, b, t)G(zms, b, t)}$$

where *(xm, ym, zm)* [μm] are the coordinates of the point where the temperature is measured, *(xs, ys, zs)* [μm] is the location of the heat source[2], and

$$\begin{aligned}
R &= RO\frac{\rho C_p \Delta}{\zeta} \\
AXP &= \frac{a/2 + (xm - xs)}{2\sqrt{D}} \\
AXN &= \frac{a/2 - (xm - xs)}{2\sqrt{D}} \\
AYP &= \frac{b/2 + (ym - ys)}{2\sqrt{D}} \\
AYN &= \frac{b/2 - (ym - ys)}{2\sqrt{D}} \\
AZP &= \frac{c/2 + (zm - zs)}{2\sqrt{D}} \\
AZN &= \frac{c/2 - (zm - zs)}{2\sqrt{D}}
\end{aligned}$$

[2]The coordinate system is assumed to coincide with the heat source principal axis.

$$G(xms, a, t) \;=\; \frac{1}{2}\left\{erf\left(AXP/\sqrt{t}\right) + erf\left(AXN/\sqrt{t}\right)\right\}.$$

In these expressions, a, b, *and* c are the dimensions of the parallelepiped (heat source) and Δ is its volume, $D = \kappa_s/\rho_s c_p$ is silicon's thermal diffusivity, and ζ is a factor between 1 and 2 that models the effect of the silicon dioxide. These parameters are defined using the *.MODEL MNAME hsmp* card as explained in Section 8.3.5.

For computational speed, the error function, *erf(x)*, is approximated in iETSIM by a second-order polynomial as follows

$$erf(x, x_o) \approx \begin{cases} \frac{x(2x_o - x)}{{x_o}^2}, & 0 \le x \le x_o \\ 1.0, & x_o \le x \\ -erf(x, x_o), & x < 0\,. \end{cases}$$

This approximation has a continuous first derivative required to avoid convergence problems. In iETSIM, x_o is set to 1.7.

The two special elements introduced in this section, namely, the power monitor and the time-dependent resistor, are generally used in an integrator configuration as depicted in Fig. 8.1 to implement the closed-form solution of the *3D* heat diffusion equation given in Chapter 7 by (see also [43])

$$T(t) = T_o + \frac{\zeta}{\rho C_p \Delta} \int_o^t G(x, a, \tau) G(y, b, \tau) G(z, b, \tau) d\tau\,.$$

In this circuit, the value of the capacitor can be scaled down provided that the value of *RO* is scaled up by the same factor. This can be used to make matrix entries more even. A suitable choice for the scale factor is for example $1.0E-6$. *Note* that the power monitor ***should be connected*** in such a way that any device power dissipation produces an increase of temperature (output voltage). *Note* that this scheme is also appropriate to implement the temperature response at a point due to superposition of different heat sources at different locations. For this, the user has to provide the corresponding power monitors and time-dependent resistors and use the *current summing* property of the integrator input node.

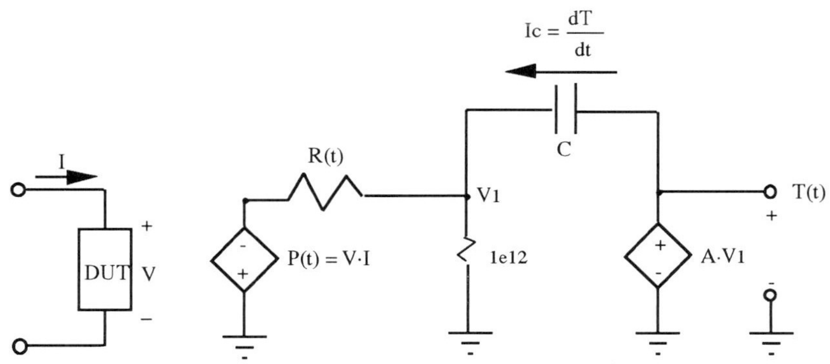

Figure 8.1 Arrangement of the special elements in an integrator circuit that implements the closed-form solution of the *3D* heat diffusion equation.

8.3.4 Semiconductor devices

Junction diodes

The general form is

DXXXXX N+ N- NTJ NTD MNAME IDEV

Examples:

D1 1 0 0 0 dm
ls 1 2 1e-9
ds 2 3 0 0 dm ls
d3 5 7 2 3 dm

N+ and N- are the anode and cathode of the diode, respectively. NTJ and NTD are the circuit nodes where the temperature readings for the junction and at the edge of the depletion region are taken (the outputs of the corresponding integrators). MNAME is the name of the model card for the diode. IDEV is the name of an inductor or voltage source connected in *series* with the diode so that the diode current can be monitored as required by the avalanche breakdown algorithm. The inductor or voltage source *should be* connected with the diode such that its terminal current is positive when the diode is forward biased.

If NTJ and NTD are set to 0, the diode parameters are held fixed, and evaluated at the value of the room temperature whose default can be overridden with the *.OPTIONS* card. If NTJ is nonzero and NTD is 0, the temperature dynamics for both impact ionization and saturation currents are controlled by NTJ. If both NTJ and NTD are nonzero, the impact ionization is evaluated at the temperature reading in NTJ, while the temperature reading in NTD is used to determine the value of the saturation current and the thermal voltage $v_t = kT/q$.

If IDEV is not specified, the diode model reduces to the ideal diode model. The avalanche breakdown effects are incorporated into the model only when a valid IDEV name is provided.

This way of handling parameters in the diode card allows the user to automatically select the type of model desired (electrical model, electrical model with avalanche breakdown, or electrothermal model) independently for each diode in the circuit.

Bipolar transistor

The general form is

QXXXXX C B E S NTCJ NTB MNAME IDEV

Examples:

Q1 9 8 0 0 0 0 qn
vct 1 2 dc 0.0
Qs 2 3 0 0 21 0 qn1 vct
q3 5 1 2 0 0 21 22 qn2

Here *C, B, E, and S* are the collector, base, emitter and substrate nodes, respectively. NTCJ and NTB are the circuit nodes from which the temperature readings for the collector junction and at the neutral-base region are taken (the outputs of the corresponding integrators). MNAME is the name of the model card associated with this device. IDEV is the name of an inductor or voltage source connected in *series* with the collector terminal so that the diode current can be monitored as required by the avalanche breakdown algorithm. The inductor or voltage source *should be* connected to the transistor such that their terminal current is positive when the collector current is positive.

If NTCJ and NTB are set to 0, the transistor model parameters are held fixed, and evaluated at the value of the value of the room temperature whose default can be changed with the *.OPTIONS* card. If NTCJ is nonzero and NTB is 0, the temperature dynamics for both collector impact ionization and saturation currents are controlled only by NTCJ. If both NTCJ and NTB are nonzero, the impact ionization is evaluated at the temperature reading in NTCJ, while the temperature reading in NTB is used to determine the value of the saturation current and the thermal voltage $v_t = kT/q$.

If IDEV is not specified, the bipolar model reduces to the Gummel-Poon model. Only when a valid IDEV name is provided are the collector avalanche breakdown effects incorporated into the model.

This way of handling parameters in the bipolar transistor card allows the user to automatically select the type of model desired (G-P model, G-P model with collector avalanche breakdown, or electrothermal model) for each bipolar transistor in the circuit.

MOS transistor

The general form is

MXXXXX D G S B NT MNAME

Examples:

m1 1 0 0 0 0 Mm

mx 5 7 2 0 35 mn2

Here *D, G, S, and B* are the drain, gate, source, and bulk nodes, respectively. *NT* is the circuit node where the temperature reading is taken. MNAME is the model card name associated with this transistor.

If *NT* is 0, no temperature coupling is used. When *NT* is a valid circuit node number, its "voltage" becomes the device temperature at which the parameters are computed.

8.3.5 Device Models

To preserve compatibility with SPICE model cards, iETSIM uses the following general form

.MODEL MNAME MTYPE < *model parameters* >

Examples

```
.model DM diode
.model QM npn BVcb=-12 br=0.2 gmin=1e-12
+ ikf = 1.0e-3 ikr =1.0e-2
.model TMOD hsmp a=62.5 b=0.4 c=0.35
+ tlin=800 zeta=1.75
```

where MNAME is any valid alphanumeric name that does not contain any delimiters, MTYPE can be *diode, npn, pnp, nmos, pmos or hsmp. [model parameters]* is a set of model parameter keywords each followed by a numeric field separated by an equal sign and possibly blank spaces. In the current implementation of iETSIM, all device models are purely dc.

Diode model parameters

The diode model is comprised of the ideal diode equation with temperature-dependent saturation current and temperature-dependent avalanche breakdown effects using a generalized Miller's multiplication factor (see Chapter 7 for details). Valid model parameters in the diode model card are

Name	Description	Default	Units
IS	saturation current	1.0e-12	[A]
XTI	IS temperature exponent	2.25	
BV	breakdown voltage at room temperature	-10	[V]
a1	$BV(T) = BV$ $+ (T - RoomT) * (a1 + a2(T - RoomT))$	-0.0011	$[V/^oK]$
a2		0.0	$[V/^oK^2]$
n0	Miller exponent for the multiplication factor M at room temperature	4.0	
n1	$n(T) = n0 + n1 * (T - RoomT)$	0.0	$[1/^oK]$
TNOM	parameter measurement temperature	27.0	$[^oC]$
gmin	parasitic conductance to aid convergence	1e-12	$[1/\Omega]$

Bipolar transistor model parameters

The bipolar model implemented in iETSIM is based on the dc Gummel-Poon model of SPICE3. This model has been enhanced in iETSIM to incorporate collector avalanche breakdown effects and a fully coupled electrothermal description capability. Valid model parameters in the bipolar transistor model card are

Name	Description	Default	Units
IS	transport saturation current	1.0e-15	[A]
XTI	IS temperature exponent	2.25	
BVcb	base-collector breakdown voltage at room temperature	-10	[V]
a1	$BVcb(T) = BVcb + (T - RoomT) * (a1 + a2(T - RoomT))$	-0.0011	$[V/^oK]$
a2		0.0	$[V/^oK^2]$
n0	Miller exponent for the multiplication factor M at room temperature	4.0	
n1	$n(T) = n0 + n1 * (T - RoomT)$	0.0	$[1/^oK]$
BF	ideal maximum forward beta	100	
NF	forward current emission coefficient	1.0	
VAF	forward Early voltage	∞	[V]
IKF	forward "knee" current	∞	[A]
ISE	B-E leakage saturation current	1.0e-15	[A]
NE	B-E leakage emission coefficient	2.0	
BR	ideal maximum reverse beta	4.0	
NR	reverse current emission coefficient	1.0	
VAR	reverse Early voltage	∞	[V]
IKR	reverse "knee" current	∞	[A]
ISC	B-C leakage saturation current	1.0e-15	[A]
NC	B-C leakage emission coefficient	2.0	
RB	zero bias base resistance	0.0	[Ω]
IRB	current where base resistance falls halfway to its minimum value	∞	[A]
RBM	minimum base resistance at high currents	RB	[Ω]
RE	emitter resistance	0.0	[Ω]
RC	collector resistance	0.0	[Ω]
TNOM	parameter measurement temperature	27.0	$[^oC]$
gmin	parasitic conductance to aid convergence	1e-12	[1/Ω]

MOS transistor model parameters

The MOS model implemented in iETSIM is based on the dc level-2 model of SPICE3. In iETSIM, this model has been expanded to account for substrate-current generation and to provide a fully coupled electrothermal description capability. Since iETSIM was initially targeted to simulation of I/O protection structures, the MOS model implemented does not have the drain and source diodes present in the SPICE model. Actually, in iETSIM, the user can incorporate an accurate description of the parasitic bipolar effect by placing in "parallel" a bipolar transistor model whenever it is desired. Valid model parameters in the MOS-2 transistor model card are

Name	Description	Default	Units
VTO	zero bias threshold voltage at room temperature	1.0	[V]
KP	transconductance parameter	2.0e-5	[A]
GAMMA	bulk threshold parameter	0.0	$[V^{1/2}]$
PHI	surface potential (ϕ)	0.6	[V]
LAMBDA	channel-length modulation	0.0	[1/V]
RD	drain resistance	0.0	[Ω]
RS	source resistance	0.0	[Ω]
TOX	oxide thickness	1.0e-5	$[cm]$
NSUB	substrate doping	0.0	$[1/cm^3]$
NSS	surface state density	0.0	$[1/cm^2]$
NFS	fast surface state density	0.0	$[1/cm^2]$
XJ	junction depth	0.0	[cm]
LD	lateral diffusion	0.0	[cm]
AI	multiplicative prefactor of the ionization current	9.0e6	[1/cm]
BI	critical sat. voltage in the exponent of the ionization current	1.7e6	[V/cm]
OPE	optical-phonon energy	0.063	[eV]
LA300	energy mean free path at 300 ^oK	10.45e-7	[cm]
TPG	type of gate material +1 opposite to substrate -1 same as substrate 0 aluminum gate	1.0	
UO	surface mobility at room temperature	600.0	$[cm^2/Vs]$
UCRIT	critical field for mobility degradation	1.0e4	[V/cm]
UEXP	critical field exponent in mobility degradation	0.0	
VMAX	maximum drift velocity of carriers	0.0	[cm/s]
NEFF	total channel charge coefficient	1.0	
DELTA	width effect on threshold voltage	0.0	
TNOM	parameter measurement temperature	27.0	$[^oC]$
gmin	parasitic conductance to aid convergence	1e-12	[1/Ω]

Temperature model parameters

The temperature model card define the substrate's temperature dependent thermal parameters, the temperature value, and the dimensions of the parallelepiped that encloses a particular heat source (see Chapter 7 for details). Valid model parameters in the temperature model card are

Name	Description	Default	Units
K	Thermal conductivity @ 300 ^oK	1.5486	$[W/cm -^o C]$
NK	$\kappa(T) = K(TLIN/300)^{NK}$	2.25	
RHOCP	Density times heat capacity @ 300 ^oK	1.574	$[J/cm^3 -^o C]$
NRCP	$\rho c_p(T) = RHOCP(TLIN/300)^{NRCP}$	0.1	
TLIN		800.0	$[^oK]$
A	Heat source dimensions $a \times b \times c$	100.0	$[\mu m]$
B		0.4	$[\mu m]$
C		0.3	$[\mu m]$
ZETA	Empirical factor to account for SiO_2	2.0	

8.3.6 Types of analysis

iETSIM can perform either a steady-state (dc) analysis or transient analysis. By default, iETSIM performs an operating point analysis if the input file does not request either DC or TRAN.

The general format of a *DC* card is

.DC SNAME START STOP STEP

where SNAME is a valid independent source name whose value is swept from START to STOP in steps of value STEP.

The general format of a *.TRAN* card is

.TRAN TSTART TSTOP TSTEP

In this case iETSIM will perform transient simulation starting at TSTART and stopping at TSTOP. If a fixed time step is used, TSTEP is the simulation time step. For a variable time step, TSTEP is used to determine the minimum and maximum time steps using the corresponding default or user-provided time-step factors.

8.3.7 Output format

When iETSIM is requested to perform an operating point analysis, it prints the values of all node voltages and current variables. In this case no *.print* card should be specified.

If iETSIM is requested to perform a dc or transient analysis, a *.print* card should be included with the following general format

.PRINT TYPE [node voltages or current variables]

where type is either *DC or TRAN*. Node voltages are specified as "V(N)" where "N" is a valid circuit node number. Current variables are specified as "I(NAME)" where "NAME" is the name of a circuit inductor or a voltage source.

Examples:

```
.print dc v(2) v(3)
.print tran v(2) i(vct) i(LX) v(10)
```

An input deck may have more than one *.print* card provided that the total number of requested outputs is not greater than 10. (This limit can be increased by recompiling the program after changing the MNOV definition in the header file *etsim.h.*)

The default output format in iETSIM is tabular. The default can be overridden if the user includes a *.XGRAPH* card in the input deck. In this case, iETSIM will generate a file that can be plotted using the plotting package XGRAPH [49]. If the *.XGRAPH* is included, an output file name should be specified in the command line.

8.3.8 Options and Debugging

The generic options card format is *.OPTIONS [ONAME = VAL]*. The following options are available:

Name	Description	Default
ITLIM	set the iteration limit	15
LTE	local truncation error to control the time step	1.0e-3
MINTSF	factor used to determine the minimum time step from TSTEP	1e-3
DX	relative tolerance for convergence	1e-9
MAXPTS	maximum number of output points	350
TNOM	room temperature value	25.0 [^oC]

When debugging new models, the user can include the *.DEBUG* card in the input file forcing iETSIM to print the circuit matrix and rhs vector before solving the linear system.

8.4 LOW TEMPERATURE THERMOMETER EXAMPLE

In this example, the electronic thermometer shown in Fig. 8.2 was simulated to determine its range of linearity. A printout of the iETSIM's input deck for this example is included for reference. iETSIM simulation results are shown in Fig. 8.3 indicating that this low-temperature thermometer loses linearity beyond 330 ^oK.

```
This deck checks the linear range of an electronic thermometer
*
.model qb npn is=8e-14 bf=200 vaf=130 br=2 xtb=-1.5 xti=2.25
.model qn npn is=5e-15 bf=200 vaf=130 br=2 xtb=-1.5
.model qp pnp is=2e-15 bf=50 vaf=50 br=4 bvcb=-10
+ xtb=-1.5 xti=2.25
*
q1 5 5 6 20 0 qb
q2 4 3 2 20 0 qp
q3 2 2 1 20 0 qp
q4 4 5 0 20 0 qn
q5 3 4 5 20 0 qn
q6 3 2 1 20 0 qp
q7 7 2 1 20 0 qp
q8 8 3 7 20 0 qp
r1 6 0 5000
r2 8 0 221.0e3
r3 2 4 5e6
vcc 1 0 dc 4.0
vtemp 20 0 dc 0.0
*
* Determine the circuit performance from -73 to 87 [oC].
* Default room temperature is TNOM = 300 [oK].
*
.dc vtemp -100 60 10
.print dc v(3) v(7) v(8)
.xgraph
*
* ITLIM was set high since no good initial guess is being provided
.options itlim=100 dx=1e-9
*
.end
```

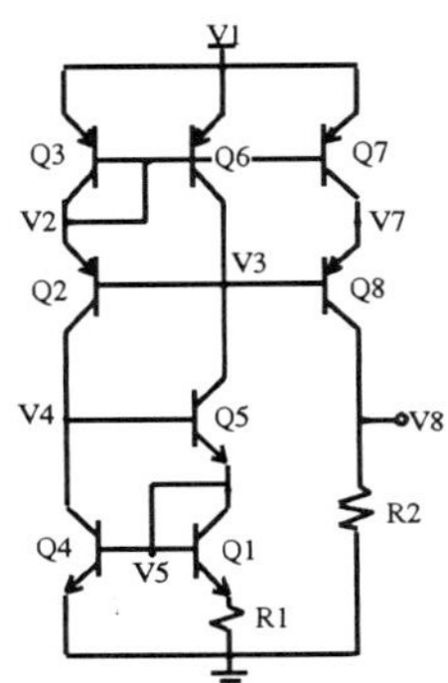

Figure 8.2 iETSIM input file used to simulate the low-temperature electronic thermometer schematically shown on the right.

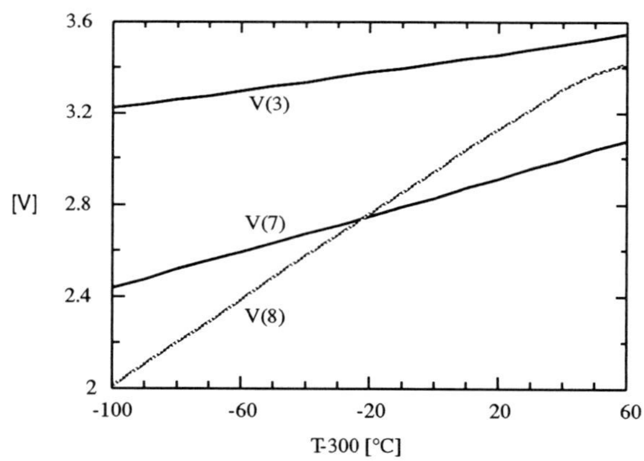

Figure 8.3 Simulation results for the low-temperature electronic thermometer example.

9 SUMMARY AND FUTURE RESEARCH

9.1 SUMMARY

An industry failure analysis database was used to evaluate the relative impact of EOS phenomena on device failure across the different product families and to identify the dominant EOS-related failure mechanisms affecting different products. A set of different devices was selected as test vehicles for this analysis. A test procedure was specified and used to consistently reproduce the dominant failure mechanisms observed in the field for the selected set of devices. Two kinds of unipolar stresses were used square-pulse testing and ESD-HBM. Stressing a number of units for each selected device type allowed us to determine the power– and current-to-failure profiles for individual devices. Detailed failure analysis results indicated that the specified test procedure consistently reproduced the given set of dominant failure mechanisms. The differences seen in the measured power profiles were qualitatively explained by failure analysis in terms of the the failure location, the layout style, and the stress level. Qualitative correlation was found between the layout style and the device's ability to sustain EOS/ESD stress as measured by the power profiles. It was shown that good IC ESD-HBM performance does not guarantee adequate EOS robustness. The power-to-failure profile was introduced to evaluate design changes of I/O protection structures in view of EOS hardness. The results in Chapter 2 suggested that power-to-failure profiles can be used to set the design goals for I/O protection against EOS and ESD and could be used as a relative measure of EOS hardness of I/O protection circuits.

The sensitivity of the power profiles to the major thermal parameters of the die-package system was investigated. To this end, the nonlinear mixed *2D-1D* thermal simulator iTSIM was developed. A proper set of boundary conditions was introduced to achieve computational efficiency when dealing with large-scale EOS thermal problems. Simulation results show that the curvature of the power profile is determined mainly by the defect geometry, defect location and thermal characteristics of the package. Simulation results suggest that the effect of the packaging material on the power profile is not dominant for times-to-failure shorter than 10 ms. The curvature of the profile is not greatly affected by the thickness of the passivation oxide, and for ceramic packages, an increase of the oxide thickness causes a slight increase of the time-to-failure for a given stress power level (upward shift of the profile). As expected, it was also observed that the peak temperature value decreases with the increase of defect distance from the Si-SiO_2 interface. Results suggest that the mixed 2D-1D thermal simulation provides a good approximation to the 3D problem and correctly predict the trends of the power profiles with respect to the various variables considered in the study.

Experiments to determine ESD-HBM and EOS failure thresholds in terms of power-to-failure and current-to-failure profiles were performed on a typical I/O protection structure used in advanced CMOS processes. The performance was measured for various values of major layout parameters such as device width and contact-to-gate spacings. Device-level electrothermal simulations were done using TMA-MEDICI [97] to explain the observed failure mechanisms. In spite of the fact that these simulations were *2D* in nature, they provided a qualitative explanation of the *3D* phenomena and correctly predicted the experimental trends of the EOS failure thresholds. Based on the simulation results, a source contact placement scheme was developed for the protection nMOS device [21]. This scheme is expected to improve the EOS/ESD performance of typical I/O protection structures used in grounded substrate CMOS ICs [21]. Simulation results suggest the presence of very steep temperature gradients within the device and indicate that it is not possible to define the onset of second breakdown in terms of a unique temperature value. Simulation results in Chapter 4 suggest that the current vs. time-to-failure ($I_f - t_f$) relationship should be used as a measure of device hardness against EOS. When the EOS can be modeled as a constant current source, the device's ability to handle the stress can then be measured by the value of t_f.

Thermal instability of an electrically stressed circuit or device was shown to be the result of either thermally induced negative differential resistance (NDR) in resistive regions, or junction second breakdown. Furthermore, it was shown that under typical ESD/EOS stress events (transient in nature), the critical temperature at which thermal instability takes place is dependent on the current stress level. Thermally induced NDR takes place in semiconductor resistors when the thermal carrier generation offsets the mobility reduction; it can be a dominant effect in semiconductor resistive regions with low doping levels and exposed to nonisothermal transient conditions. A general criterion for thermal instability of semiconductor resistors is introduced in terms of the sheet resistance and the temperature profile. In reverse-biased semiconductor junctions under transient nonisothermal conditions, second breakdown is shown to happen at the time when thermal carrier generation becomes high enough to offset the effects of the mobility degradation and the reduction of the impact ionization rates. Under these circumstances, the time for the onset of second breakdown is shown to depend on the device's geometry and the level of power dissipation.

To provide an appropriate CAD framework for studying electrothermal effects at the circuit design level, new algorithms and device models were introduced that enable accurate circuit simulation of networks with devices operated up to avalanche breakdown conditions under both isothermal and nonisothermal conditions. These models and algorithms were implemented in the electrothermal circuit simulator iETSIM that self-consistently solves for the device temperatures and electrical circuit variables using the Modified Nodal Analysis approach. iETSIM uses a closed-form solution of the $3D$ heat diffusion equation to determine the device temperatures as a function of the device power dissipation. SPICE models for the bipolar and MOS transistors were enhanced in iETSIM to incorporate both the avalanche breakdown phenomena and the temperature dynamics of the model electrical parameters. Currently, iETSIM can describe the behaviour of ESD/EOS protection circuits up to the onset of second breakdown without lost of convergence; iETSIM can also be used to study circuit performance sensitivity to device self-heating effects.

Analysis of the experimental results, the theoretical temperature model, and the newly proposed criteria for thermal instability of semiconductor elements led to the development of a design concept for improved ESD/EOS protection. This design concept that calls for minimum device self-heating has been implemented in an area-efficient protection circuit that is expected to provide robust ESD/EOS protection for integrated circuits fabricated in advanced n-well CMOS processes [29].

9.2 FUTURE RESEARCH

To make device-level electrothermal simulators suitable for EOS studies, it is necessary to implement appropriate thermal boundary conditions such as those introduced in Chapter 3. This is required for accurate and computationally efficient simulation of EOS events lasting longer than a few microseconds since the heat wavefront can propagate far away from the region where the heat generation is taking place.

Most of the results presented here deal with MOS technologies for which the ESD/EOS I/O protection is generally achieved by the parasitic bipolar transistor of an MOS device operating in the snapback domain. Similar studies have to be conducted on protection devices for bipolar and BiCMOS technologies, particularly regarding the impact of layout and process parameters on the failure of vertical bipolar transistors.

The newly proposed protection schemes for CMOS processes [21, 29] and implemented in the MOSIS test chip have to be evaluated in terms of the ESD-HBM and square-pulse testing for EOS through power– and current-to-failure profiles.

In the electrothermal analysis of device failure under EOS/ESD events, one assumes that the expressions describing silicon's physical parameter dependencies on lattice temperature hold beyond the temperature range for which they were originally derived. Thus, it is necessary to investigate the validity of these assumptions particularly for the models of the mobility and the impact ionization coefficients.

Currently, the device models implemented in iETSIM are purely dc in nature. The user has to lump capacitive effects into linear capacitors connected to the device terminals in the circuit description file. For accurate analysis of dynamic circuit performance sensitivity of analog or digital circuits with respect to device self-heating, it is necessary to incorporate the nonlinear charge storage effects into the device models as is done in SPICE for transient analysis. The iETSIM code can also be extended to handle subcircuit description as is done in SPICE-like circuit simulation environments.

BIBLIOGRAPHY

[1] A. E. Ruehli, editor. *Circuit Analysis, Simulation and Design*, volume 3, part 1, chapter 2, pages 45–70. North-Holland, Amsterdam, 1986.

[2] M. Adler. Accurate calculations of the forward drop and power dissipation in thyristors. *IEEE Transactions on Electron Devices*, ED-25(1):16–22, January 1978.

[3] A. Amerasekera, L. van Roozendaal, J. Abrahalden, J. Bruines, and L.Sevant. An analysis of low voltage ESD damage in advanced CMOS processes. In *Proceedings of the EOS/ESD Symposium*, pages 143–152, 1990.

[4] Ajith Amerasekera, Amitava Chatterjee, and M. Chang. Prediction of ESD robustness in a process using 2-D device simulations. In *Proceedings of the IEEE International Reliability Physics Symposium*, pages 161–167, March 1993.

[5] Ajith Amerasekera, L. Roozendaal, J. Bruines, and F. Kuper. Characterization and modeling of second breakdown in nMOST's for the extraction of ESD-related process parameters. *IEEE Transactions on Electron Devices*, ED-38(9):2161–2168, September 1991.

[6] P. Antognetti and G. Massobrio. *Semiconductor Device Modeling with SPICE*. McGraw Hill, New York, 1988.

[7] M. Ash. Semiconductor junction non-linear failure power thresholds: Wunsch-Bell revisited. In *Proceedings of the EOS/ESD Symposium*, pages 122–127, September 1983.

[8] H. S. Carslaw and J. Jaeger. *Conduction of Heat in Solids*. Oxford Press, Oxford UK, 1959.

[9] B. Carvajal, R. Cline, and B. Anderson. A successful HBM ESD protection circuit for micron and sub-micron level CMOS. In *Proceedings of the EOS/ESD Symposium*, pages 234–242, 1992.

[10] M. Cavone, M. Muschitiello, R. Rivoir, and M. Stucchi. Analysis of current distribution in NPN transistors during ESD stress by means of gated emission microscopy. *Unpublished*, 1992.

[11] M. Chang, J. Chern, and P. Yang. Electrothermal device simulations using path tracing algorithms. *Unpublished*, 1993.

[12] R. Chapman, C. Wei, D. Bell, S. Aur, G. Brown, and R. Haken. 0.5 microm CMOS for high performance at 3.3V. In *IEEE International Electron Devices Meeting Technical Digest*, pages 52–55, 1988.

[13] A. Chatterjee and T. Polgreen. A low-voltage triggering SCR for on-chip ESD protection at output and input pads. *IEEE Electron Device Letters*, EDL-12:21–22, January 1991.

[14] R. Chemelli and L. DeChiaro. The characterization and control of leading edge transients from Human-Body-Model ESD simulators. In *Proceedings of the EOS/ESD Symposium*, pages 155–162, September 1985.

[15] K. Chen. The effects of interconnect process and snapback voltage on the ESD failure threshold of nMOS transistors. *IEEE Transactions on Electron Devices*, ED-35(12):2140–2150, 1988.

[16] K. Chen, G. Giles, and D. Scott. Electrostatic discharge protection for one micron CMOS devices and circuits. In *IEEE International Electron Devices Meeting Technical Digest*, pages 484–487, 1986.

[17] R. Consiglio. Private Comunication. LSI Logic Corporation, 1993.

[18] C. Crowell and S. Sze. Temperature dependence of avalanche multiplication in semiconductors. *Applied Physics Letters*, 9(6):242–244, September 1966.

[19] G.T. Dangelmayer and E.S. Jesby. Employee training for successful ESD control. In *Proceedings of the EOS/ESD Symposium*, 1985.

[20] S. Daniel and D. Krieger. Process and design optimization for advanced CMOS I/O ESD protection devices. In *Proceedings of the EOS/ESD Symposium*, pages 206–213, September 1990.

[21] C. Díaz, C. Duvvury, and S. Kang. Source Contact Placement for Efficient ESD/EOS Protection in Grounded-Substrate MOS ICs. US Patent filed, 1993. Patent Activity TI-17832.

[22] C. Díaz, C. Duvvury, and S. Kang. Thermal failure simulation for electrical overstress in semiconductor devices. In *IEEE International Symposium on Circuits and Systems*, 1993.

[23] C. Díaz, C. Duvvury, S. M. Kang, and L. Wagner. Electrical overstress (EOS) power profiles: a guideline to qualify EOS hardness of semiconductor devices. In *Proceedings of the EOS/ESD Symposium*, pages 88–94, 1992.

[24] C. Díaz and S. Kang. Modeling and Simulation of Electrical Overstress Failures in ICs. Technical Report TI-3Q-91, Texas Instruments Incorporated, 1991.

[25] C. Díaz and S. Kang. Modeling and Simulation of Electrical Overstress Failures in ICs. Technical Report TI-3Q-91, Texas Instruments Incorporated, 1991.

[26] C. Díaz and S. Kang. New algorithms for circuit simulation of device breakdown. *IEEE Transactions on Computer Aided Design*, CAD-11(11):1344–1354, November 1992.

[27] C. Díaz and S. Kang. Circuit-level electrothermal simulation of electrical overstress failures in advanced MOS I/O protection devices. *IEEE Transactions on Computer Aided Design*, 13(4):482–493, April 1994.

[28] C. Díaz, S. Kang, and C. Duvvury. Electrothermal simulation of electrical overstress in advanced nMOS ESD I/O protection devices. In *IEEE International Electron Devices Meeting Technical Digest*, pages 899–902, 1993.

[29] C. Díaz, S. Kang, and C. Duvvury. New ESD/EOS Protection Structures for Integrated Circuits Fabricated in Advanced n-well CMOS Processes. US Patent filed, 1993. Patent Activity TI-17833.

[30] C. Díaz, S. Kang, and C. Duvvury. Thermal failure simulation for electrical overstress in semiconductor devices. *IEEE Transactions on Electron Devices*, 41(3):359–366, March 1994.

[31] C. Díaz and G. Motley. Bi-modal triggering for LVSCR ESD protection devices . In *Proceedings of the EOS/ESD Symposium*, 1994.

[32] C. Duvvury. ESD reliability for advanced CMOS technologies. In *Electron Devices and Materials Symposium*, Taiwan, 1990.

[33] C. Duvvury and A. Amerasekera. ESD: a pervasive reliability concern for IC technologies. *Proceedings of the IEEE*, 81(5):690–702, May 1993.

[34] C. Duvvury and C. Díaz. Dynamic gate coupling of nMOS for efficient output ESD protection. In *Proceedings of the IEEE International Reliability Physics Symposium*, pages 141–150, March 1992.

[35] C. Duvvury, C. Díaz, and T. Haddock. Achieving uniform nMOS device power distribution for sub-micron ESD reliability. In *IEEE International Electron Devices Meeting Technical Digest*, pages 131–134, December 1992.

[36] C. Duvvury and R. Rountree. A synthesis of ESD input protection scheme. *Journal of Electrostatics*, 29:1–19, 1992.

[37] C. Duvvury, R. Rountree, D. Baglee, and R. McPhee. ESD protection reliability in one micron CMOS technologies. In *Proceedings of the IEEE International Reliability Physics Symposium*, pages 199–205, 1986.

[38] C. Duvvury, R. Rountree, R. McPhee, D. Baglee, A. Hyslop, and L. White. ESD design considerations for ULSI. In *Proceedings of the EOS/ESD Symposium*, pages 45–48, September 1985.

[39] C. Duvvury, R. Rountree, H. Stiegler, T. Polgreen, and D. Corum. ESD phenomena in graded junction devices. In *Proceedings of the IEEE International Reliability Physics Symposium*, pages 71–76, 1989.

[40] C. Duvvury, R. Rountree, and L. White. A summary of most effective electrostatic discharge protection circuits for MOS memories and their observed failure modes. In *Proceedings of the EOS/ESD Symposium*, pages 181–184, September 1983.

[41] C. Duvvury and R.N. Rountree. Output ESD techniques for advanced CMOS processes. In *Proceedings of the EOS/ESD Symposium*, 1988.

[42] V. Dwyer, A. Franklin, and D. Campbell. ESD thermal failure in semiconductor devices. *IEEE Transactions on Electron Devices*, ED-37:2345–2354, November 1990.

[43] V. Dwyer, A. Franklin, and D. Campbell. Thermal failure in semiconductor devices. *Solid State Electronics*, 33:553–560, May 1990.

[44] S. Ghandhi. *Semiconductor Power Devices.* Wiley-InterScience, New York, 1977.

[45] P. Gray and D. Hamilton. Analysis of electrothermal integrated circuits. *IEEE Journal of Solid State Circuits*, SC-6(1):8–14, February 1971.

[46] Paul R. Gray. *Analysis and Design of Analog Integrated Circuits.* Wiley, New York, 2nd edition, 1984.

[47] T. Green and W. Denson. A Review of EOS/ESD field failures in military equipment. In *Proceedings of the EOS/ESD Symposium*, pages 7–13, September 1988.

[48] X. Gui, P. Webb, and G. Gao. Use of the three-dimensional TLM method in the thermal simulation and design of semiconductor devices. *IEEE Transactions on Electron Devices*, ED-39(6):1295–1301, June 1992.

[49] David Harrison. *XGRAPH User's Guide.* University of California at Berkeley, 1989.

[50] M. Hassan, A. Khondker, and H. Domingos. Effects of graded collector doping on current mode second breakdown in bipolar transistors. In *Proceedings of the EOS/ESD Symposium*, pages 70–76, September 1988.

[51] H. Hill and D.P. Renaud. ESD in semiconductor wafer processing. In *Proceedings of the EOS/ESD Symposium*, 1985.

[52] F. Hsu, R. Miller, and C. Hu. A simplified model for short-channel MOSFET characteristics in the breakdown mode. *IEEE Transactions on Electron Devices*, ED-30:571–576, June 1983.

[53] C. Hu, S. Tam, F. Hsu, P. Ko, T. Chan, and K. Terril. Hot-electron-induced MOSFET degradation – model, monitor, and improvement. *IEEE Transactions on Electron Devices*, ED-32:375–384, February 1985.

[54] J. Huang, H. Chen, and J. Kueng. Modeling of output snapback characteristics in n-channel SOI MOSFETs. *IEEE Transactions on Electron Devices*, ED-39(5):1170–1178, May 1992.

[55] D. Klinger, Y. Nakada, and M. Menendez, editors. *AT&T Reliability Manual.* Van Nostrand Reinhold, New York, 1st edition, 1990.

[56] B. Krabbenborg, R. Beltman, P. Wolbert, and T. Mouthaan. Physics of electrothermal effects in ESD protection devices. In *Proceedings of the EOS/ESD Symposium*, pages 98–103, 1991.

[57] K. Kundert and A. Sangiovanni-Vicentelli. Sparse Matrix Routines. Technical Report , University of California at Berkeley, 1989.

[58] N. Kusnezov and J. Smith. Modeling of electrical overstress in silicon devices. In *Proceedings of the EOS/ESD Symposium*, pages 132–138, 1981.

[59] M. Latif and P. Bryant. Network analysis approach to multidimensional modeling of transistors including thermal effects. *IEEE Transactions on Computer Aided Design*, 1(2):94–100, April 1982.

[60] S. Laux and F. Gaensslen. A study of channel avalanche breakdown in scaled n-MOSFETs. *IEEE Transactions on Electron Devices*, ED-34(5):1066–1073, 1987.

[61] C. Lee, A. Palisoc, and Y. Min. Thermal analysis of integrated circuit devices and packages. *IEEE Transactions on Components, Hybrids, and Manufacturing Technology*, 12(4):700–709, December 1989.

[62] R. Lindsted and R. Surty. Steady-state junction temperatures in semiconductor chips. *IEEE Transactions on Electron Devices*, ED-19(1), January 1972.

[63] T. Maloney and N. Khurana. Transmission line pulsing techniques for circuit modeling of ESD phenomena. In *Proceedings of the EOS/ESD Symposium*, pages 49–54, September 1985.

[64] K. Mayaram, J. Chern, L. Arledge, and P. Yang. Electro-thermal simulation tools for analysis and design of ESD protection devices. In *IEEE International Electron Devices Meeting Technical Digest*, pages 909–912, 1991.

[65] O. McAteer. An effective ESD awareness training program. In *Proceedings of the EOS/ESD Symposium*, 1980.

[66] O. McAteer, R.E. Twist, and R.C. Walker. Latent ESD failures. In *Proceedings of the EOS/ESD Symposium*, 1982.

[67] J. McPhee. The effects of backdriving integrated circuits - an accurate electrothermal model. In *Proceedings of the IEEE International Test Conference*, pages 518–522, 1985.

[68] R. McPhee, C. Duvvury, R. Rountree, and H. Domingos. Thick oxide ESD performance under process variations. In *Proceedings of the EOS/ESD Symposium*, 1986.

[69] R. Merrill and E. Issaq. ESD design methodology. In *Proceedings of the EOS/ESD Symposium*, pages 233–237, 1993.

[70] M. Middendorf and T. Hanksen. Observed physical defects and failure analysis of EOS/ESD on semiconductor devices. In *International Symposium for Test and Failure Analysis*, pages 205–213, 1984.

[71] MIL-STD-883C. Electrostatic Discharge Sensitivity Classification. Technical Report Notice 8, DOD, March 1989.

[72] S. Miller. Ionization rates for holes and electrons in silicon. *Physical Review*, 105(4):1246–1249, February 1957.

[73] W. Orvis, J. Yee, G. Khanaha, and D. Lair. Theoretical Modeling of EMP Induced Breakdown Effects in Semiconductor Devices. Technical Report UCID-19900, Lawrence Livermore Laboratory, June 1991.

[74] A. Palella and H. Domingos. A design methodology for ESD protection networks. In *Proceedings of the EOS/ESD Symposium*, pages 24–39, September 1985.

[75] D. Pierce and D. Durgin. An overview of electrical overstress effects on semiconductor devices. In *Proceedings of the EOS/ESD Symposium*, pages 120–131, September 1981.

[76] D. Pierce, W. Shiley, B. Mulcahy, K. Wagner, and M. Wunder. Electrical overstress testing of a 256K UVEPROM to rectangular and double exponential pulses. In *Proceedings of the EOS/ESD Symposium*, pages 137–146, September 1988.

[77] T. Polgreen and A. Chaterjee. Improving the ESD failure threshold of silicided nMOS output transistors by ensuring uniform current flow. *IEEE Transactions on Electron Devices*, ED-39(2):379–388, February 1992.

[78] G. Poon and J. Meckwood. Modeling avalanche effect in integral charge control model. *IEEE Transactions on Electron Devices*, ED-19(1):90–97, January 1972.

[79] K. Poulton, K. Knudsen, J. Corcoran, K. Wang, R. Pierson, R. Nubling, and M. Chang. Thermal design and simulation of bipolar integrated circuits. *IEEE Journal of Solid State Circuits*, SC-27(10):1379–1387, October 1992.

[80] M. Reisch. On bistable behaviour and open-base breakdown of bipolar transistors in the avalanche regime - modeling and applications. *IEEE Transactions on Electron Devices*, ED-39(6):1398–1409, June 1992.

[81] Robert Renninger, M. Jon, D. Ling, T. Diep, and T. Welsher. A field-induced charged-device model simulator. In *Proceedings of the EOS/ESD Symposium*, pages 59–71, September 1989.

[82] L. Roozendaal, A. Amerasekera, P. Bos, W. Baelde, F. Bontekoe, P. Kersten, E. Korma, P. Roomers, P. Krys, U. Weber, and P. Ashby. Standard ESD testing of integrated circuits. In *Proceedings of the EOS/ESD Symposium*, pages 119–130, September 1990.

[83] R. Rountree. ESD protection for CMOS circuits: issues and solutions. In *IEEE International Electron Devices Meeting Technical Digest*, pages 580–583, 1988.

[84] R. Rountree, C. Duvvury, T. Maki, and H. Stiegler. A process tolerant input protection circuit for advanced CMOS processes. In *Proceedings of the EOS/ESD Symposium*, pages 201–205, 1988.

[85] R.N. Rountree and C.L. Hutchins. NMOS protection circuitry. *IEEE Transactions on Electron Devices*, ED-32:910–917, May 1985.

[86] A. Rubalcava and W. Roesch. Lack of latent and cumulative ESD effects on MESFET-based GaAs IC's. In *Proceedings of the EOS/ESD Symposium*, 1988.

[87] Harry Schafft. Second breakdown – a comprehensive review. *Proceedings of the IEEE*, 55(8):1272–1285, August 1967.

[88] A. Schütz, S. Selberherr, and H. Pötzl. A two-dimensional model for the avalanche effect in MOS transistors. *Solid State Electronics*, 25(3):177–183, 1982.

[89] D. Scott, G. Giles, and J. Hall. A lumped element model for simulation of ESD failures in silicided devices. In *Proceedings of the EOS/ESD Symposium*, pages 41–47, September 1986.

[90] S. Selberherr. *Analysis and Simulation of Semiconductor Devices*. Springer-Verlag, New York, 1984.

[91] S. Shabde, G. Simmons, A. Baluni, and R. Back. Snapback induced gate dielectric breakdown in graded junction nMOS transistors. In *Proceedings of the IEEE International Reliability Physics Symposium*, pages 168–175, 1984.

[92] M. Storm. Heat conduction in simple metals. *Journal of Applied Physics*, 22(7):940–950, July 1951.

[93] E. Sun, J. Moll, J. Berger, and B. Alders. Breakdown mechanism in short-channel MOS transistors. In *IEEE International Electron Devices Meeting Technical Digest*, pages 478–482, 1978.

[94] R. Sunshine and M. Lampert. Second-breakdown phenomena in avalanching silicon-on-sapphire diodes. *IEEE Transactions on Electron Devices*, ED-19:873–885, July 1972.

[95] S. M. Sze. *Physics of Semiconductor Devices*. Wiley, New York, 2nd edition, 1981.

[96] D. Tasca. Pulse power failure modes in semiconductors. *IEEE Transactions on Nuclear Science*, 17:346–372, December 1970.

[97] Technology Modeling Associates, Inc., Palo Alto, California. *MEDICI, Two Dimensional Device Simulation Program*, 1992.

[98] University of California at Berkeley, California. *Spice3e2 User's Manual*, 1991.

[99] G. Wachutka. Rigorous thermodynamic treatment of heat generation and conduction in semiconductor device modeling. *IEEE Transactions on Computer Aided Design*, 9(11):1141–1149, November 1990.

[100] D. Wunsch and R. Bell. Determination of threshold failure levels of semiconductor diodes and transistors due to pulse power voltages. *IEEE Transactions on Nuclear Science*, 15:244–259, 1968.

[101] A. Yang and S. Kang. iSMILE: a novel circuit simulation program with emphasis on new device model development. In *26th ACM/IEEE Design Automation Conference*, pages 630–633, 1989.

INDEX

ABOUT THE AUTHORS

Carlos H. Díaz received the B.S. and M.S. degrees in electrical engineering, and the B.S. degree in physics from Universidad de Los Andes, Bogotá - Colombia, in 1983, 1985, and 1984, respectively. He received the Ph.D. degree in electrical engineering from the University of Illinois at Urbana-Champaign in May 1993. During 1985 he was engaged in the design of an automatic production system for a concrete plant. In 1986, he joined the faculty of the Universidad de Los Andes as a professor in the Department of Electrical Engineering. From 1988 to 1989 he was the Undersecretary of Information Sciences in the Presidencia de la República de Colombia. He was a research assistant at the Coordinated Science Laboratory, University of Illinois at Urbana-Champaign from January 1990 to May 1993. He held summer positions in IC reliability at National Semiconductor Corporation, Santa Clara, CA, and Texas Instruments Incorporated, Dallas, TX, in 1990 and 1991-92, respectively. Currently, he is a member of the technical staff in the Integrated Circuits Business Division R&D Center, Hewlett-Packard Company, Palo Alto, CA. His research interests include semiconductor device modeling, device and circuit simulation, and IC reliability. He has published over 15 papers in technical journals and conferences. He has five U.S. patent applications currently at various stages of the patent filing process. He received the SRC Inventor Recognition Award in 1993. He is a member of the IEEE and the Asociación Colombiana de Ingenieros Eléctricos, Mecánicos y Afines (ACIEM).

Sung Mo (Steve) Kang received the Ph.D. degree in electrical engineering from the University of California at Berkeley in 1975. Until 1985, he was with AT&T Bell Laboratories at Holmdel, Murray Hill, and also served as a faculty member at Rutgers University. At AT&T Bell Laboratories, he worked on large-scale telecommunications network planning, design and manufacturing of WE32000-series VLSI microprocessor chips and peripheral chips. In 1985, he joined the University of Illinois at Urbana-Champaign where he is Professor of Electrical and Computer Engineering, Computer Science and Research Professor at the Coordinated Science Laboratory, and Associate Director of the NSF Engineering Research Center for Compound Semiconductor Microelectronics. He is also an Associate in the Center for Advanced Study at the University of

Illinois. He has served on the editorial boards of *IEEE Transactions on Circuits and Systems, IEEE Circuits and Devices Magazine, IEEE Design & Test of Computers, International Journal of Circuit and Theory and Applications, and Circuits, Systems and Signal Processing Journal.* He was special guest editor for three special issues of *IEEE Design & Test of Computers* and *International Journal of Circuit Theory and Applications.* He has served on the program committees of DAC, ICCAD, ICCD, ISCAS, MWSCAS and international workshops. His research interests include modeling of semiconductor devices, VLSI CAD and design methodologies, optimization techniques for performance, reliability and manufacturability, and optoelectronic circuits and systems. He has been a consultant for the semiconductor industry and delivered lectures at several international workshops on VLSI design and CAD. He has received best paper awards (1979, 1987), IEEE Computer Society Meritorious Service Award (1990), IEEE Darlington Award (1993) and SRC Inventor Recognition Award (1993). Dr. Kang is the Founding Editor of *IEEE Transactions on VLSI Systems* and has served as Administrative Vice President, Secretary and Treasurer, AdCom member and 1991 President of IEEE Circuits and Systems Society.

Charvaka Duvvury received the B.S.E.E. from the University of Houston and the M.S.E.E. and Ph.D. in Materials Science from the University of Toledo. After working as a postdoctoral Fellow in Physics at the University of Alberta in Canada, he joined Texas Instruments in 1977 as a Design Engineer in the MOS Memory Group in Houston. He worked on DRAM design, transistor modeling, and reliability issues that included hot carriers and ESD from 1977-1988. In 1988 he joined the Semiconductor Process and Design Center in Dallas, Texas, where he is currently a Senior Member of the Technical Staff working on ESD reliability research in advanced CMOS. He has published over 35 papers in technical journals and conferences and holds eight patents. He has taught short courses on ESD and hot carrier offered by the University of California at Berkeley as well as made several international presentations on ESD. Recently he contributed towards a chapter for a book on hot carriers published by Van Nostrand Reinhold. He is a member of the IEEE, Eta Kappa Nu, and Sigma XI.